A Gateway to Number Theory

Applying the Power of Algebraic Curves

AMS/MAA | DOLCIANI MATHEMATICAL EXPOSITIONS

VOL 57

A Gateway to Number Theory

Applying the Power of Algebraic Curves

Keith Kendig

Providence, Rhode Island

Dolciani Mathematical Expositions Editorial Board
C. Ray Rosentrater, Editor

Priscilla S. Bremser   C. L. Frenzen   Tom Richmond
Alfred M. Dahma   Kim Regnier Jongerius   Ayse A. Sahin
Matt Delong   Katharine Ott   Daniel E. Steffy

Cover image: Glenn Nagel/Shutterstock.com

2020 *Mathematics Subject Classification.* Primary 11-XX, 11-01, 11D04, 11D09, 11D25, 11G05, 14H45, 14H52.

For additional information and updates on this book, visit
www.ams.org/bookpages/dol-57

Library of Congress Cataloging-in-Publication Data
Names: Kendig, Keith, 1938- author.
Title: A gateway to number theory : applying the power of algebraic curves / Keith Kendig.
Description: Providence : American Mathematical Society, [2021] | Series: Dolciani mathematical expositions; volume 57 | Includes bibliographical references and index.
Identifiers: LCCN 2020052775 | ISBN 9781470456221 (paperback) | ISBN 9781470465025 (ebook)
Subjects: LCSH: Diophantine equations. | Curves, Algebraic. | Number theory. | AMS: Number theory. | Number theory – Instructional exposition (textbooks, tutorial papers, etc.). | Number theory – Diophantine equations – Linear equations. | Number theory – Diophantine equations – Quadratic and bilinear equations. | Number theory – Diophantine equations – Cubic and quartic equations. | Number theory – Arithmetic algebraic geometry (Diophantine geometry) – Elliptic curves over global fields. | Algebraic geometry – Curves – Special curves and curves of low genus. | Algebraic geometry – Curves – Elliptic curves.
Classification: LCC QA242 .K46 2021 | DDC 512.7/4–dc23
LC record available at https://lccn.loc.gov/2020052775

Copying and reprinting. Individual readers of this publication, and nonprofit libraries acting for them, are permitted to make fair use of the material, such as to copy select pages for use in teaching or research. Permission is granted to quote brief passages from this publication in reviews, provided the customary acknowledgment of the source is given.

Republication, systematic copying, or multiple reproduction of any material in this publication is permitted only under license from the American Mathematical Society. Requests for permission to reuse portions of AMS publication content are handled by the Copyright Clearance Center. For more information, please visit www.ams.org/publications/pubpermissions.

Send requests for translation rights and licensed reprints to reprint-permission@ams.org.

© 2021 by the author. All rights reserved.
Printed in the United States of America.

∞ The paper used in this book is acid-free and falls within the guidelines
established to ensure permanence and durability.
Visit the AMS home page at https://www.ams.org/

10 9 8 7 6 5 4 3 2 1   26 25 24 23 22 21

To Manjul Bhargava

Contents

Preface   xi

1 A Marriage for the Ages   1
    1.1 Mathematics and the Hawaiian Islands   1
    1.2 A Model   2
    1.3 Appearance of the Rational Unit Circle, Part I   3
    1.4 Appearance of the Rational Unit Circle, Part II   8
    1.5 What About Other Rational Circles?   9
    1.6 A Parametric Formula for the Unit Circle in $\mathbb{Q}^2$ 12
    1.7 More General Quadratic Problems   15
    1.8 Conics   21
    1.9 Our Method Also Works in Reverse   26
    1.10 Surveying All Solutions   29
    1.11 The Discriminant   30
    1.12 Finding a Rational Point   33
    1.13 What About Degree 1?   34

2 Viewing the Whole Algebraic Curve   37
    2.1 An Ancient Indian Parable   37
    2.2 Viewing Algebraic Curves   37
    2.3 Doing Math at Infinity!   48
    2.4 A More Symmetric Model of $\mathbb{P}^2(\mathbb{R})$ 53

3 Entering the World of Elliptic Curves   55
    3.1 Curves of Degree 3   55
    3.2 What Is an Elliptic Curve?   57
    3.3 Why Irreducible?   59
    3.4 Smoothness   60
    3.5 Weierstrass Forms for Elliptic Curves   64

3.6	The Discriminant Again	68
3.7	Our Method Still Works	72
3.8	Examples of Connecting the Dots	78
3.9	Mordell's Theorem	80

4 Every Elliptic Curve Is a Group! — 81
4.1	Finite Groups Within an Elliptic Curve	82
4.2	Finitely Generated Abelian Groups	90
4.3	Rank	92
4.4	Mazur's Theorem	94
4.5	A Gallery of Positive Rank Elliptic Curves	97
4.6	How Many Curves?	100
4.7	Finding Generators	101

5 A Million-Dollar Challenge — 105
5.1	Breaking Up a Task into Many Smaller Ones	105
5.2	The Birch and Swinnerton-Dyer Conjecture	110
5.3	The Notion of Expected Rank	119
5.4	Expected Rank of a Random Elliptic Curve	120
5.5	The Tale of Average Rank	122
5.6	Rank Results Without GRH or BSD	127
5.7	About Manjul Bhargava	130

6 Every Real Elliptic Curve Lives in a Donut — 137
6.1	Complex Curves	137
6.2	Complex Numbers Enlighten	138
6.3	Plotting a Complex Circle	139
6.4	Plotting a Complex Elliptic Curve	142
6.5	Subgroups and Cosets	145
6.6	Elliptic Curves with No Oval	152

7 The Genus — 155
7.1	A Few Preliminaries	155
7.2	Examples	156
7.3	The Genus Formula	158
7.4	The Genus vs. Number Theory	159
7.5	The Curious Story of Plane vs. Fancy Curves	162

Contents ix

8 In Conclusion . . . 165
    8.1   Degree 1 165
    8.2   Degree 2 166
    8.3   Degree 3 168

A What Is a Smooth Complex Curve? 171

B Algebraic Curves in the Disk Model 181
    A Teachable Picture 184
    Some Basic Definitions 186
    Further Examples 189

C Some Code for This Book's Programs 193

Bibliography 201

Index 205

Preface

This book is about a major breakthrough in an important part of number theory and how algebraic curves play a central role in this revolution. Many problems that once posed a challenge for experienced mathematicians have become virtually routine to solve. This book is an informal and leisurely introduction to understanding this revolution and how to apply its methods to solve a large class of what are called "Diophantine equations." In general, solving such equations means finding integer solutions to polynomial equations with integer coefficients. In many of the most important cases, the polynomial terms all have the same degree, and then we can apply a powerful method that works like a blowtorch: Translate the problem to one in geometry where intuition is a potent aid and solve it there. Then translate back to the world of integers to see the final solution.

Our approach is concrete with lots of pictures and solved numeric problems. The power and simplicity of the new method makes a large swath of formerly difficult problems easy to solve. In fact, you'll be able to devise nontrivial number theory problems on your own and solve them using the method.

The background needed for the early chapters of this book is mainly high school math. Some familiarity with Mathematica and Maple will be helpful, and for those wishing to use freeware, GeoGebra fills the bill admirably. Using any of these powerful tools requires some coding, and Appendix C supplies code or pseudocode for those wanting it. In the last half of the book we encounter some basic concepts covered in a typical undergraduate math major, mostly from beginning group theory and linear algebra. The book also includes sketches of some enlightening theory. To read these parts, a basic course in complex variables and one in topology should suffice.

A little Q&A between writer and potential reader will help elucidate the overall design of the book:

- *What kind of number theory problems are we talking about?* They are problems in three integer unknowns. For example, find all integer solutions to
$$a^2 + b^2 = c^2$$
or to
$$3a^2 - 5c^2 = 11b^2$$
or to really crazy ones like
$$-91b^2 + 37c^2 + 84a^2 = 0.$$
You, or a computer, could doubtless find some specific solutions, *but finding all of them?* It turns out that either there are no solutions at all or an infinite number of them. The approach helps you decide which is which, and in the infinite case, the method provides a way of finding them all.

- *What kind of geometry are we talking about?* We're talking about curves defined by an equation $p(x, y) = 0$, where $p(x, y)$ is a polynomial having integer coefficients and degree ≤ 3. That means lines, circles, ellipses, parabolas, hyperbolas, and a wide variety of cubic curves.

- *What do we do with these curves?* We drain away most of the points in the plane, leaving only those having rational numbers for both coordinates. What's left of the curve constitutes precisely the geometric version of all solutions to the problem. We then translate back to the world of integers. (A quick aside: A rational number is any number that can be written as a fraction or ratio of integers such as $\frac{2}{1}$ or $\frac{5}{10}$. The nonfractions 2 and 0.5 are rational numbers since they can be rewritten as fractions. Notice that "rational" contains the word "ratio.")

- *But how do we translate to geometry?* We divide the equation through by a power of a, b, or c to get rid of it. For example, divide $a^2 + b^2 = c^2$ by c^2 to get $x^2 + y^2 = 1$, where x and y are the rational numbers $\frac{a}{c}$ and $\frac{b}{c}$. It's $x^2 + y^2 = 1$ that defines our curve — in this case, a circle. After draining

Preface

away all the nonrational points, we're staring at *the geometric solutions to the problem.*

- Impressive, but how do we actually get specific solutions? And from what you said, there must be an infinite number of them! Still looking at $a^2 + b^2 = c^2$ as a representative example, pick a point such as $(-1, 0)$ on the corresponding circle $x^2 + y^2 = 1$. Through this point, pass a line having rational slope and solve for the other point where the line meets the circle. This involves only high school math, but it's powerful. Just make up any rational number you want for the slope, and you get another solution. As we'll see in the book, the rational number $\frac{1}{2}$ gives $\left(\frac{3}{5}, \frac{4}{5}\right)$. To get integers, just multiply $\left(\frac{3}{5}, \frac{4}{5}\right)$ through by the denominator. For example,

$$\left(\frac{3}{5}\right)^2 + \left(\frac{4}{5}\right)^2 = 1$$

becomes

$$3^2 + 4^2 = 5^2.$$

So $\frac{1}{2}$ gives the famous right triangle $(3, 4, 5)$. You can make up fancier rationals and get corresponding right triangles. For example, $\frac{7}{5}$ gives the right triangle $(24, 70, 74)$ — a more unusual example.

- So you just multiply through to change rational numbers into integers? That's how you translate from geometry to number theory? That's it. The only restriction on the original a, b, c problem is that when you translate to a curve, you get a polynomial $p(x, y)$ whose locus is a curve. That means when you divide by, say, c^2 in $a^2 + b^2 = c^2$, there is no c left over after writing $x = \frac{a}{c}$ and $y = \frac{b}{c}$. That always works provided the terms involving a, b, c all have the same degree. We call such an equation "homogeneous." That's true of all the examples mentioned above, but it wouldn't be true for something like $a^2 + b^2 = c$, for example. So in fancier language, this book is about

> solving homogeneous Diophantine equations of degree ≤ 3.

- So you get every possible integer solution using this method? Very close to that! Here's how to get all solutions to any degree-two homogeneous Diophantine equation in integer variables a, b, c: Let m run through the values $\mathbb{Q}\cup\{\infty\}$ and for each such value, find a "primitive" solution — a triple of integers (a, b, c) where a, b, c have no common factors other than ± 1. If the triple you have isn't primitive, you can make it so by dividing the triple through by the greatest common divisor of (a, b, c). Then up to sign, any solution to the homogeneous Diophantine equation is obtained by multiplying that primitive solution by some integer to get (an, bn, cn). For the problem $a^2 + b^2 = c^2$, for example, the slope $\frac{1}{2}$ leads to $(3, 4, 5)$ which is primitive, so up to signs, every solution corresponding to $\frac{1}{2}$ is

$$(\pm 3n, \pm 4n, \pm 5n) \ (n \in \mathbb{Z}),$$

where the $\pm$ are taken independently from coordinate to coordinate. And that slightly strange case of slope $\frac{7}{5}$? It led to $(24, 70, 74)$. This triple has 2 as a common factor, so $(12, 35, 37)$ is primitive, and every solution of $a^2 + b^2 = c^2$ corresponding to that slope is

$$(\pm 12n, \pm 35n, \pm 37n) \ (n \in \mathbb{Z}).$$

I want to express my great gratitude to those who have helped make this book a reality. Its original inspiration and much of its content came from Manjul Bhargava's beautiful Hedrick Lectures on this subject given during the Mathematical Association of America's summer meeting in 2011. It is my pleasure to dedicate this book to him. (See the biographical sketch about him on pp. 130–135.) I am also greatly indebted to Don Albers, past Acquisitions Editor of the MAA, who was both constant cheerleader and wise counsel in writing this book. Stephen Kennedy took over after Don's retirement and has likewise given me much helpful advice. Special thanks are due Harriet Pollatsek, Editor of the Dolciani Series, as well as her successor Ray Rosentrater. The entire Dolciani Board, appearing on p. iv, has been a model of thoroughness; the book has been significantly improved, thanks to their conscientious work. I would also like to express appreciation in a slightly different direction. Just as a concert pianist owes a debt of gratitude to the years of work and ingenuity that went into the many improvements resulting in a magnificent instrument, I too owe a debt of gratitude to the designers of the powerful and reliable software that became my constant companions as I wrote this book. At the top of the list has to be PCTeX used in writing the text. As for

Preface

the figures, Adobe's Illustrator and Photoshop were indispensable to me. On the mathematical side of the ledger, Maple, Mathematica, and GeoAlgebra were all a great help. Kudos to the many talented developers of such beautiful software!

Keith Kendig
Chagrin Falls, OH
March 2021

1

A Marriage for the Ages

1.1 Mathematics and the Hawaiian Islands

We begin our journey with a surprising analogy between mathematics and the Hawaiian Islands. Each can be thought of as an archipelago. In mathematics, as in the Hawaiian Islands, some islands are large — areas such as classical algebra, geometry, topology, number theory, and so on. There are smaller ones, too, say Martingale methods in statistics. In both math and the Hawaiian Islands, the islands are not actually separate but have connections that are not always obvious. From the viewpoint of someone in an airplane looking down, Hawaii looks like a collection of land masses separate from each other. But in reality the airplane passenger is just seeing the tops of a mostly submerged mountain range, and the whole mountain range isn't seen because the water obscures the view. If we could pull a plug and let the water drain away, we would gradually see the whole mountain range come into view, and we'd discover that the islands are in fact connected. The mathematical analog of water is ignorance, as it clouds our ability to see the whole. Pulling the plug in this case corresponds to decreasing the level of ignorance, and as that happens, mathematicians discover connections — often unsuspected — between areas of math that were previously assumed to be separate.

Discovering connections between seemingly unrelated ideas has often been a basic feature of significant mathematics. For example, old-fashioned geometry and old-fashioned algebra each developed separately for centuries, but with the birth of analytic geometry and its coordinate systems, two big islands got connected: Points became ordered pairs, and lines, circles, and conics became equations. Geometric theorems could be translated into algebra and proved using algebra's great power. The connection was a two-way road, meaning that algebra could be translated into geometry. For example, fancy polynomial equations eventually became translated into algebraic curves, giving rise to the mathematical branch of algebraic geometry. This marriage between geometry and algebra was destined to be one for the ages.

This book is about another marriage for the ages. Let's identify the actors.

Definition 1.1.1.

> A *plane algebraic curve* is the set of solutions in the real plane $\mathbb{R}^2$ of a polynomial equation $p(x, y) = 0$. If the coefficients of $p(x, y)$ are rational — or equally well, integers — the curve is called *rational*. In this book we assume any curve is rational unless stated otherwise.

Algebraic curves slink and curve around in the plane. Rational curves are intimately connected to discrete points forming the essence of number theory, and in this book we attempt to lower the water level to reveal basic connections between number theory (which is discrete), rational curves (which usually slink around), as well as some topology and even statistics.

1.2 A Model

A great idea in mathematics often has a model containing the basic core of that idea. For us, our model is one of the most famous equations in mathematics: $a^n + b^n = c^n$, where we look for solutions in positive integers a, b, c, and n. It turns out that there is one simple operation that has defined much of the recent progress in number theory. It has proved to be one of the most important steps in bringing together the two big fields of algebraic curves and number theory. Applied to our model, here's that one simple operation:

> Divide both sides of $a^n + b^n = c^n$ by c^n to get $x^n + y^n = 1$, so that x and y are rational.

What happens to these pictures when we let all nonrational points drain away? That is, when we plot $x^n + y^n = 1$ in $\mathbb{Q}^2$, where $\mathbb{Q}$ stands for the set of all rational numbers? It turns out that for $n = 2$, the unit circle still looks like a unit circle. For all other n, the transformation is entirely different. Instead of a plot in the real plane looking like a curve, when n is even we see in $\mathbb{Q}^2$ only the four points $(\pm 1, 0)$ and $(0, \pm 1)$. Things get even more sparse for odd n, because then only $(1, 0)$ and $(0, 1)$ survive, a consequence of Fermat's Last Theorem:

Theorem 1.2.1 (Fermat's Last Theorem).

> For integers $n > 2$, the equation $a^n + b^n = c^n$ cannot be solved using only nonzero integers a, b, and c.

Comment 1.2.2. Fermat stated his famous "Last Theorem" in 1637. He claimed he'd proved it but that the margins of his notebook were too small to contain his proof, a claim that virtually no mathematician believes today. Fermat's conjecture stubbornly resisted seemingly countless attempts at proof and became arguably the most famous unsolved problem in mathematics. Finally, after seven years of intense work, Andrew Wiles cracked it in 1994, 357 years after its original statement. His proof today stands as one of the outstanding success stories in mathematics. ◇

1.3 Appearance of the Rational Unit Circle, Part I

Everyone knows what the circle $x^2 + y^2 = 1$ looks like in the real plane $\mathbb{R}^2$. Remarkably, when we eliminate nearly all the original points and look at the solution set in only the rational plane $\mathbb{Q}^2$, what we see still looks like a circle. But what does $x^2 + y^2 = 1$ *actually* look like in $\mathbb{Q}^2$? For example, suppose we magnify this set a billion or trillion times. Will we find missing arcs of positive length — "gaps" — in it? Or, no matter how much we magnify, will such gaps never appear? We find the answer in this section.

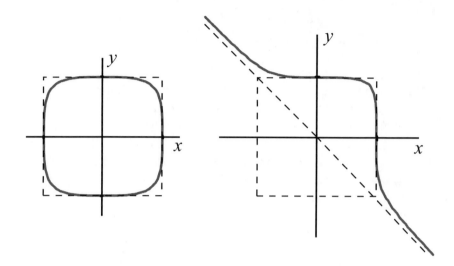

Figure 1.1. Plots in $\mathbb{R}^2$ of "Fermat curves" $x^n + y^n = 1$ illustrating their appearance for even versus odd n. The curve on the left is of $n = 4$, and on the right, $n = 5$. As n increases, each curve hugs the dashed lines more closely. When $n = 100$, for example, the curve on the left would look just like a square. When magnified enough, however, we'd see that the corners are actually rounded.

Let's begin by remembering that $x^2 + y^2 = 1$ came from the Pythagorean Theorem $a^2 + b^2 = c^2$ applied to positive integers a, b, and c, where we've divided each side by c^2 to get

$$\left(\frac{a}{c}\right)^2 + \left(\frac{b}{c}\right)^2 = 1.$$

As before, x and y in $x^2 + y^2 = 1$ are the rational numbers $\frac{a}{c}$ and $\frac{b}{c}$. Some of these (a, b, c) right triangles are famous, like $(3, 4, 5)$, $(6, 8, 10)$, $(5, 12, 13)$, and $(8, 15, 17)$. There are also not-so-famous ones, like $(11, 60, 61)$, and ones that hardly anybody knows or cares about, like $(4{,}691,\ 6{,}480,\ 8{,}161)$. The point here is that when we divide any of these triples by the integer c, we get rational numbers x and y satisfying $x^2 + y^2 = 1$. That is, we get points on the circle $x^2 + y^2 = 1$ in $\mathbb{Q}^2$. But under tremendous magnification, does what we see through our microscope still look like a very tiny circular arc greatly magnified?

1.3. Appearance of the Rational Unit Circle, Part I

First steps toward an answer go back at least 3,800 years to the Old Babylonian Empire in Mesopotamia. Evidence is strong that even then people knew some form of the Pythagorean Theorem. Construction of right angles was important for laying out buildings, plots of land, and countless other projects from large to small.

Figure 1.2. Plimpton 322. Source: Wikimedia Commons.

In a major stroke of luck, in the early 1900s, a clay cuneiform tablet was discovered in southern Iraq, and around 1922 a publisher from New York — George Plimpton — bought the tablet from an archaeological dealer for what was even then a very reasonable $10. This tablet turned out to open a rare window on some of the mathematics those ancient Mesopotamians knew. Among each of the tablet's 15 rows are integers corresponding to a, b, c in an (a, b, c) right triangle. What's so remarkable is that the tablet contains triples far removed from simple ones like $(3, 4, 5)$, $(6, 8, 10)$, $(5, 12, 13)$, or $(8, 15, 17)$ which could have been discovered by trial alone. Here, for example, are four triplets from four rows in the tablet:

- $(319, \ 360, \ 481)$,
- $(799, \ 960, \ 1{,}249)$,
- $(2{,}291, \ 2{,}700, \ 3{,}541)$,
- $(12{,}709, \ 13{,}500, \ 18{,}541)$.

The remarkable size of these triples has lead many historians to believe those ancients must have had some sort of formula or algorithm for generating them, perhaps even that they had figured out the gist of Euclid's algorithm. Here's one form of his algorithm:

Choose two positive integers p and q. Then, as in Figure 1.3, these integers generate the Pythagorean triple (a, b, c), with

$$\mathbf{a} = |\mathbf{p}^2 - \mathbf{q}^2|,$$

$$\mathbf{b} = \mathbf{2pq},$$

$$\mathbf{c} = \mathbf{p}^2 + \mathbf{q}^2.$$

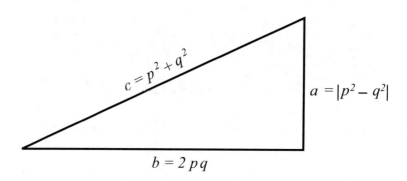

Figure 1.3. Euclid's formula.

These equations are sometimes called Euclid's formula since Euclid includes a proof in his *Elements*. From Figure 1.3 we see that

$$a^2 + b^2 = (p^4 - 2p^2q^2 + q^4) + (4p^2q^2) = p^4 + 2p^2q^2 + q^4 = c^2.$$

Comment 1.3.1. Much of ancient Greek mathematics was based on lengths, areas, and volumes. Negative numbers were not meaningfully addressed until the Indian mathematician Brahmagupta used them in the 7th century A.D. to represent debts. For us, however, p, q in Figure 1.3 can be chosen from the set of all integers $\mathbb{Z}$ rather than the positive ones — the natural numbers $\mathbb{N}$ — meaning the absolute sign in $|p^2 - q^2|$ is unnecessary. ◇

1.3. Appearance of the Rational Unit Circle, Part I

With the abundance of integer pairs, it's easy to generate a huge number of points $(\frac{a}{c}, \frac{b}{c})$ on the unit circle. However, this alone isn't enough to insure that in the rational plane $\mathbb{Q}^2$, the unit circle centered at the origin has no missing arcs when viewed at arbitrarily high magnifications. As an example, consider the line $y = \pi x$. If x is rational, then y isn't, so although in $\mathbb{R}^2$ the line intersects this unit circle in a point P, P isn't in $\mathbb{Q}^2$. To establish that there is no gap around P in this rational circle, we need to find a rational number m arbitrarily close to π so that $y = mx$ intersects the rational circle.

Here's the gist of a general argument. From Figure 1.3 we have

$$m = \frac{p^2 - q^2}{2pq} = \frac{1}{2}\left[\frac{p}{q} - \frac{q}{p}\right].$$

For a moment, consider $\frac{p}{q}$ to be any real number, writing $\frac{1}{2}\left[\frac{p}{q} - \frac{q}{p}\right]$ as $\frac{1}{2}\left[\frac{r}{1} - \frac{1}{r}\right]$. Now for a little magic: Let r be $\pi + \sqrt{\pi^2 + 1}$. A bit of algebra shows that $\frac{1}{2}\left[\frac{r}{1} - \frac{1}{r}\right]$ simplifies to exactly π. We can use this and continuity as a way to get our desired approximation by choosing $\frac{p}{q}$ to be rational and close to $\pi + \sqrt{\pi^2 + 1}$. A little thought shows that $\frac{1}{2}\left[\frac{p}{q} - \frac{q}{p}\right]$ will then be rational and close to π.

The following example shows the above argument in action. Begin by approximating

$$\pi + \sqrt{\pi^2 + 1} \in \mathbb{R}$$

to some number of decimal places — say,

$$3.14159 + \sqrt{3.14159 + 1} = 6.43850 = \frac{643{,}850}{100{,}000}.$$

Let p be the numerator 643,850 and q, the denominator 100,000. The slope

$$m = \frac{p^2 - q^2}{2pq} = \frac{1}{2}\left[\frac{p}{q} - \frac{q}{p}\right] = \frac{1}{2}\left[\frac{643{,}850}{100{,}000} - \frac{100{,}000}{643{,}850}\right]$$

computes to $m = 3.14159 \in \mathbb{Q}$, correct to six figures. The more places of the approximation to $\pi + \sqrt{\pi^2 + 1}$ we take, the closer our computed m will be to π.

Notice that in all this, any irrational number can be used in place of π, and we can go a bit further and say that P above can be any point of the circle not in $\mathbb{Q}^2$. One can then argue that all such P form a set in the circle with no gaps — that is, there's no missing arc of positive length in the set. In official language, this no-gaps locus in $\mathbb{Q}^2$ is *everywhere dense* in the locus in $\mathbb{R}^2$ of $x^2 + y^2 = 1$. So what we've shown above says that the rational points in the circle are everywhere dense there.

Exercise 1.3.2. Determine the fewest number of places $\pi + \sqrt{\pi^2 + 1}$ requires to get π accurate to 20 places.

1.4 Appearance of the Rational Unit Circle, Part II

The method we just met uses Euclid's formula whose roots go back to the Babylonians, some 1,500 years before Euclid. We now look at a more modern and powerful approach where, instead of a picture of the circle $x^2 + y^2 = 1$ and lines passing through the origin, we pick a rational point P on the circle and consider lines passing through P. This seemingly small change turns out to pay huge dividends. Let's see how it works.

First, pick a point P on our circle. For convenience, let P be $(-1, 0)$. Then a line L with slope m through P has equation $y = m(x + 1)$. We take m to be rational, an assumption central to our new approach. The set of points on our rational circle is tremendously thinned out compared to the entire real circle, so what are the chances that L ends up intersecting the circle in some rational point Q? For an answer, plug L's equation $y = m(x + 1)$ into $x^2 + y^2 = 1$ and see what happens. The substitution gives

$$x^2 + m^2(x+1)^2 = 1,$$

which can be expanded and rewritten as

$$(m^2 + 1)x^2 + 2m^2 x + (m^2 - 1) = 0. \tag{1.1}$$

This is a quadratic equation with rational coefficients, and it's easy to check that indeed, $x = -1$ is one root of this equation, no matter what m is. We could solve for the other root, but actually we don't need to. Instead, divide the quadratic in (1.1) by the leading coefficient to make the polynomial monic. Now in any monic quadratic, the coefficient of x is the negative sum of the two roots:

$$x^2 + ax + b = (x - r)(x - s) = x^2 + (-r - s)x + rs.$$

1.5. What About Other Rational Circles?

In our case this coefficient is rational, as is the root $x = -1$. So the other root, whatever it may be, must be rational — that is, the x-coordinate of Q is rational. But the line L's equation is $y = m(x+1)$, so the y-coordinate of Q must be rational, too. Therefore L does in fact intersect our thin set of rational points on the circle.

We know even more: Because both coordinates of any rational point of the circle are rational, substituting those coordinates into $y = m(x+1)$ means that the slope m is rational. We can even take $m = \infty$, corresponding to a vertical line through $(-1, 0)$. This intersects the rational circle in a double point consisting of the original point $(-1, 0)$ counted twice. We have now established that $\mathbb{Q} \cup \{\infty\}$, looked at as slopes of lines through $(-1, 0)$, naturally parameterizes via intersection all rational points of our circle, and the circle has no gaps in it.

Although we took our base point P to be $(-1, 0)$, any other rational point on our circle works, too. The algebra may be more complicated, but the overall logical arguments are the same.

Exercise 1.4.1. Redo the above argument using $P = (0, 1)$ instead of $(-1, 0)$.

1.5 What About Other Rational Circles?

We have shown in two different ways that the locus of $x^2 + y^2 = 1$ in $\mathbb{Q}^2$ actually does look like a circle — there are no missing arcs in the set, even of arbitrarily small positive length. But what about the loci in $\mathbb{Q}^2$ of $x^2 + y^2 = 2$ or $x^2 + y^2 = 3$ or $x^2 + y^2 = 4$, and so on? Are *they* circles in this same sense? Starting with any positive integer n and a rational point P on $x^2 + y^2 = n$, we can follow through the steps in the last section, and it appears that the locus of $x^2 + y^2 = n$ in $\mathbb{Q}^2$ does look like a circle with no missing arcs. But concrete examples are often great teachers, so let's take a look!

Example 1.5.1. For the locus of $x^2 + y^2 = 2$ we can take $(1, 1)$ as a rational base point P, since $1^2 + 1^2 = 2$. The steps in the section just above lead to a locus in $\mathbb{Q}^2$ that is dense in the full circle — that is, it has no missing arcs of positive length in it. ◇

Example 1.5.2. What about $x^2 + y^2 = 3$? In order to follow the steps in the last section to ensure that the locus of $x^2 + y^2 = 3$ in $\mathbb{Q}^2$ looks like a circle with no gaps, we need to first find a rational point P in its locus. It turns out we can try, try, and try some more, yet our desired rational point P remains elusive. One can run through many different candidates for a few hours, days, or years, even using a high-speed computer search program, yet the crucial rational point P continues to elude us. We may conclude that there simply isn't such a P, but of course all such testing doesn't constitute a proof. The next exercise pins down that there really is no rational point on the circle $x^2 + y^2 = 3$. It's essentially a parity (even versus odd) argument. ◊

Exercise 1.5.3. To show there's no rational point on the circle $x^2+y^2 = 3$, we begin by assuming that the nonzero integers a, b, and c have no common factor — if they do, just divide each of a, b, c by that factor. Now $a^2 + b^2 = 3c^2$ involves sums of squares. If c is even, say $2n$, then its square is $4n^2$. Therefore c^2 is 0 (mod 4).

(a) Show, similarly, that if c is odd, then its square is 1 (mod 4).

(b) When c is even, $3c^2$ is $3 \cdot 0 = 0$ (mod 4). Show that in this case, if there's a solution, a^2 and b^2 must both be odd or both even.

(c) Suppose in (b) that a^2, b^2, and c are all even. Then each of a, b, c has 2 as a factor. Why does this contradict our original assumption?

(d) Therefore suppose in (b) that a^2 and b^2 are both odd (and c is even). Show this means a^2 and b^2 are both 1 (mod 4), implying $a^2 + b^2$ is 2 (mod 4). Why is this impossible?

(e) Therefore if there's a solution, c must be odd. Show that means $3c^2$ is 3 (mod 4). For the left-hand side to be odd as well, show that $a^2 + b^2$ is 1 (mod 4). Therefore c can't be odd because if there's a solution, both sides must be the same (mod 4).

Conclusion: c is either even or odd, but in both cases we find that assuming a solution leads to a contradiction.

The next theorem is far more powerful and general than the parity argument above. It provides a reliable way of deciding whether or not there's a rational point P not only on the circle $x^2 + y^2 = 3$, but on any curve $Ax^2 + By^2 + C = 0$, where A, B, and C integers. It's due to the French mathematician Adrien-Marie Legendre (1752–1833).

1.5. What About Other Rational Circles?

Theorem 1.5.4 (Legendre's Criterion).

> There is a rational point on the locus of $Ax^2 + By^2 + C = 0$ exactly when we know these things about A, B, and C:
>
> - Not all of A, B, and C have the same sign. (When all of them are positive, or all are negative, the locus is empty.)
>
> - ABC is squarefree. (A number is squarefree if there are no repeated factors in the number's factorization into primes.) As an example, $2 \cdot 3 \cdot 5$ is squarefree, but $3 \cdot 3 \cdot 5$ is not.
>
> - $-AB$ is a square (mod C),
>
> - $-AC$ is a square (mod B),
>
> - $-BC$ is a square (mod A).

For a proof, see [Ireland, Propositions 17.3.1 and 17.3.2].

Let's use this powerful tool on $x^2 + y^2 = 3$. Here, $A = 1$, $B = 1$, and $C = -3$. Not all of these have the same sign, so the first hurdle is passed. Also, $ABC = -3$ which is squarefree, so the second condition is satisfied. Next, is $-AB$ a square (mod C)? We see that $-AB = -1$ and $C = -3$. Is -1 a square in $\mathbb{Z}_{-3} = \mathbb{Z}_3$? ($\mathbb{Z}_3$ means the integers (mod 3), which is the same as the 3-hour-clock with hours 0, 1, 2.) In $\mathbb{Z}_3$, -1 is 2 and the squares of $\mathbb{Z}_3$ are $0 \cdot 0 = 0$, $1 \cdot 1 = 1$, and $2 \cdot 2 = 4 = 1$. Well, well! 2 is not on the list, so Legendre's Criterion has already dealt a fatal blow — not all the requirements on $A = 1$, $B = 1$, and $C = -3$ are satisfied, so there is *no* rational point P in the locus of $x^2 + y^2 = 3$. Our suspicions are now solidly confirmed: $x^2 + y^2 = 3$ is the empty set in the rational plane! When we talked about choosing a rational point P in the locus and then all the logic used for $x^2 + y^2 = 1$ would apply, meaning that the locus would look like a circle with no gaps, that assumed there *is a rational point P to begin with*.

From this, we see that

> The locus of $x^2 + y^2 = n$ (n a positive integer) in $\mathbb{Q}^2$ is either a set looking like a circle or we see nothing at all. There is no in-between — it's all or nothing.

Exercise 1.5.5. In plotting the locus of $x^2 + y^2 = n$ in $\mathbb{Q}^2$, determine whether we see a circle or the empty set for each of $n = 4, 5, 6, 7$.

Exercise 1.5.6. For which of $n = 3, 4, 5, 6$ does the locus of $x^2 + 2y^2 = n$ in $\mathbb{Q}^2$ look like an ellipse?

Exercise 1.5.7. For which of $n = 3, 4, 5, 6$ does the locus of $x^2 - y^2 = n$ in $\mathbb{Q}^2$ look like a hyperbola?

1.6 A Parametric Formula for the Unit Circle in $\mathbb{Q}^2$

We have just shown that the lines of rational slope m through $P = (-1, 0)$ parameterize the rational points of the unit circle $x^2 + y^2 = 1$. We can derive an explicit formula for the parametrization, and this will lead to a solution (a, b, c) to $a^2 + b^2 = c^2$ corresponding to each $m \in \mathbb{Q} \cup \{\infty\}$.

To get the parametrization, argue as we did just after (1.1) on p. 8: First divide the quadratic in (1.1) by the leading coefficient to make the polynomial monic. As we noted then, in any monic quadratic the coefficient of x is the negative sum of the two roots:

$$x^2 + ax + b = (x - r)(x - s) = x^2 + (-r - s)x + rs.$$

We know one root $x_1 = -1$, so the other root must satisfy

$$\frac{2m^2}{m^2 + 1} = -(x_2 - 1).$$

Solving for x_2 gives

$$\frac{1 - m^2}{1 + m^2}.$$

Of course, one could alternatively apply the usual formula for the roots of a quadratic equation to the one in equation (1.1) and get the same answer.

To get the y-coordinate of our parametrization, substitute $\frac{1-m^2}{1+m^2}$ for x in $y = m(x + 1)$, which is the equation of the line with slope m through $P = (-1, 0)$; see Figure 1.4. Doing this and simplifying gives our desired parametrization:

$$m \to Q = \left(\frac{1 - m^2}{1 + m^2}, \frac{2m}{1 + m^2}\right). \tag{1.2}$$

1.6. A Parametric Formula for the Unit Circle in $\mathbb{Q}^2$

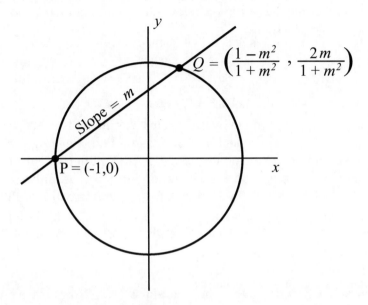

Figure 1.4. This figure shows the line $y = m(x + 1)$ through $(-1, 0)$. As $m \in \mathbb{Q}$ increases from 0 to 1, Q moves counterclockwise, covering the part of the circle in the first quadrant. As m increases from 1 to ∞, Q covers the part in the second quadrant. As m symmetrically decreases from 0 to -1, and from -1 to $-\infty$, Q moves clockwise, covering the fourth and third quadrants.

Our parametrization (1.2) leads to a solution in integers this way: Write m as a fraction $\frac{p}{q}$ of integers and substitute this directly into (1.2) to obtain explicit formulas for a, b, and c in terms of p and q. This substitution gives

$$\left(\frac{1 - \frac{p^2}{q^2}}{1 + \frac{p^2}{q^2}}, \frac{\frac{2p}{q}}{1 + \frac{p^2}{q^2}} \right),$$

which simplifies to

$$\left(\frac{-p^2 + q^2}{p^2 + q^2}, \frac{2pq}{p^2 + q^2} \right).$$

Multiplying through by p^2+q^2 gives the following solution to $a^2+b^2=c^2$ corresponding to each $m = \frac{p}{q} \in \mathbb{Q}$:

$$\begin{aligned} \mathbf{a} &= |\mathbf{p}^2 - \mathbf{q}^2|, \\ \mathbf{b} &= \mathbf{2pq}, \\ \mathbf{c} &= \mathbf{p}^2 + \mathbf{q}^2. \end{aligned} \qquad (1.3)$$

Dividing this (a, b, c) through by the greatest common divisor of a, b, and c — written $\gcd(a, b, c)$ — yields a primitive solution, meaning a solution having no common factors other than ± 1. Any solution to our problem is then obtained by multiplying each coordinate of this primitive solution by some integer n, whose sign may vary from coordinate to coordinate.

Slopes of simple fractions such as $\frac{1}{2}, \frac{1}{3}, \frac{2}{3}$, and so on yield famous right triangles, while more complicated fractions yield less famous or barely known right triangles.

Comment 1.6.1. Look again at the formulas in (1.3). Now turn to p. 6 and look at the boldfaced formulas there! These absolutely identical formulas were discovered at least 3,800 years apart by completely different methods — a remarkable testament to the timelessness of mathematics.

Example 1.6.2. With $m = \frac{p}{q} = \frac{1}{2}$, plugging $p = 1$ and $q = 2$ into (1.3) gives $(a, b, c) = (3, 4, 5)$, which is primitive. If $\frac{p}{q}$ is not in lowest terms — say, numerator and denominator share an integer factor γ — then each of a, b, and c is multiplied by γ^2. So, for example, $\frac{3}{6}$ produces $(a, b, c) = (27, 36, 45)$. ◊

Here are a few other examples.

Example 1.6.3. $m = \frac{1}{3}$ yields the triangle $(a, b, c) = (8, 6, 10)$. Since $\gcd(8, 6, 10) = 2$, a primitive solution is $(4, 3, 5)$. ◊

Example 1.6.4. $m = \frac{2}{3}$ produces the triple $(a, b, c) = (5, 12, 13)$, which is primitive. ◊

1.7. More General Quadratic Problems

Example 1.6.5. $m = \frac{1}{4}$ gives the solution $(a, b, c) = (15, 8, 17)$, and this is primitive. ◇

Example 1.6.6. $m = \frac{3}{4}$ yields the solution $(a, b, c) = (7, 24, 25)$, which is primitive. ◇

Example 1.6.7. $m = \frac{1}{5}$ produces $(a, b, c) = (24, 10, 26)$, with primitive solution $(12, 5, 13)$. ◇

Example 1.6.8. $m = \frac{2}{5}$ yields $(a, b, c) = (21, 20, 29)$. ◇

Here's a slightly more unusual choice:

Example 1.6.9. $m = \frac{7}{5}$ gives $(a, b, c) = (24, 70, 74)$. Since $\gcd(24, 70, 74) = 2$, a primitive solution is $(12, 25, 37)$. ◇

Exercise 1.6.10. Verify the results stated in Examples 1.6.3 through 1.6.9.

Exercise 1.6.11. Assume $m = \frac{p}{q}$ is written in lowest terms. Do finite rational slopes m and $\frac{1}{m}$ always yield the same ordered triple? Why or why not?

Exercise 1.6.12. Notice that choosing $p = 1$, $q = 2$ above yields the same shape of triangle as $p = 1$, $q = 3$. Similarly, $p = 2$, $q = 3$ yields the same shape of triangle as $p = 1$, $q = 5$. Are there similar pairs for the other examples above?

1.7 More General Quadratic Problems

We now outline a method for solving a huge swath of quadratic number theory problems. The three formulas in (1.3) lead to all integer solutions of $a^2 + b^2 = c^2$. Impressive though this is, in the world of quadratic number theory problems, $a^2 + b^2 = c^2$ is pretty tame and is only one of infinitely many possibilities. What about finding formulas for all integer solutions a, b, and c to

$$a^2 + b^2 = 2c^2 \text{ ?}$$

Or upping the ante and asking for formulas for
$$2a^2 + 3b^2 = 5c^2 ?$$
Or getting much wilder and asking for ones giving all integer solutions to
$$5a^2 + 8b^2 + 3ab + 11bc - 5c^2 = 0 ?$$
These seem to head toward some really serious challenges. A computer, by running through integer triples (a, b, c), may well find many specific solutions provided they exist, but what about finding *all* solutions? Or definitely establishing that there exist *no* solutions? That's quite a different matter. Notice that all these problems involve three integer variables a, b, and c, and every term has degree 2. For these problems the basic approach used above works. It's straightforward, potent, and elegant. (What more can a mathematician ask for?)

Let's make an official definition.

Definition 1.7.1. A polynomial p is *homogeneous of degree n* provided every term of the polynomial has degree n. The associated equation $p = 0$ is likewise called *homogeneous*. ◇

Now that we've had some experience in solving problems, we sharpen a bit what we said in the Preface:

> This book is about using algebraic curves over $\mathbb{Q}$ to solve homogeneous Diophantine equations of degree ≤ 3.

We've seen that when the degree is 2, then if we can find one solution, that will lead to all the rest of the solutions. To recap the method:

> Convert the homogeneous quadratic equation in integers a, b, and nonzero c to an equation in $x = \frac{a}{c}$ and $y = \frac{b}{c}$. If this equation defines a nondegenerate conic — that is, an ellipse, parabola, or hyperbola — and if there's a rational point on this conic, then any line through that point intersects the conic in a second (not necessarily distinct) point. When the line's slope is rational, the second point is also rational, and we can then go backwards from this point to an integer solution to the original equation. This leads to all its integer solutions.

1.7. More General Quadratic Problems

If there is no rational point on the curve, then the locus in $\mathbb{Q}^2$ is empty and the original number theory problem has no integer solutions. If there's even one rational point, then there are infinitely many solutions $(x, y) \in \mathbb{Q}^2$ and therefore infinitely many integer solutions to the original number theory problem. In that case, we can work out parametrizations of x and y in terms of the rational slope m, and by writing $m = \frac{p}{q}$, we can use these parametrizations to get explicit formulas in terms of p and q for a solution (a, b, c) corresponding to m.

A good way to appreciate the power of the above method is to try finding on your own a couple of solutions to the problem below. After doing that, the solution we get will take on added significance. So using our method, let's solve this problem:

> Find all integer solutions to $2a^2 + 3b^2 = 5c^2$.

Here are the steps in the solution:

- First, transform $2a^2 + 3b^2 = 5c^2$ to an equation in x and y to get an algebraic curve. Dividing through by c^2 gives $2x^2 + 3y^2 = 5$ — that is, $\frac{2}{5}x^2 + \frac{3}{5}y^2 = 1$. This defines the ellipse depicted in Figure 1.5.

- Second, find a rational point on the ellipse; $(x, y) = (1, 1)$ does it.

- Now, between the equations of line and ellipse, eliminate y, which we can do by writing the line's equation $y - 1 = m(x - 1)$ as $y = mx - m + 1$ and substituting this y for y in the ellipse equation $2x^2 + 3y^2 = 5$. We get

$$2x^2 + 3(mx - m + 1)^2 = 5.$$

By expanding and collecting terms, this can be put into the standard form $Ax^2 + Bx + C = 0$:

$$(3m^2 + 2)x^2 + 6m(1 - m)x + (3m^2 - 6m - 2) = 0.$$

Our challenge is to factor this to get its roots, which will then lead to the desired parametrization of (x, y) in terms of m. Since the line goes through $P = (1, 1)$, we already know that one factor is $(x - 1)$. To get the other factor, long divide the quadratic by $(x - 1)$. This division turns out to be straightforward and gives

$$(3m^2 + 2)x + (-3m^2 + 6m + 2).$$

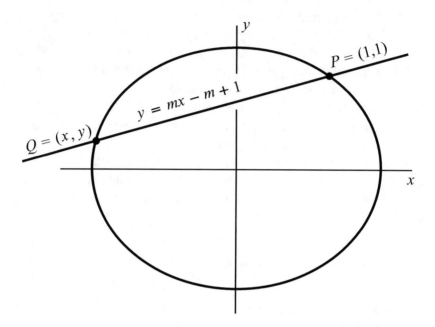

Figure 1.5. This sketch of the ellipse $2a^2 + 3b^2 = 5c^2$ shows the line $y = mx - m + 1$ through the rational point $P = (1, 1)$. As m runs through $\mathbb{Q} \cup \{\infty\}$, all the infinitely many points of the ellipse in $\mathbb{Q}^2$ are covered. When $m = \infty$, the vertical line through P intersects the ellipse in the other point $(1, -1)$. The ellipse itself has semi-major and semi-minor axes $\sqrt{\frac{5}{2}}$ and $\sqrt{\frac{5}{3}}$.

Setting this equal to zero and solving for x gives the first coordinate of our parametrization:
$$x = \frac{3m^2 - 6m - 2}{3m^2 + 2}. \tag{1.4}$$
Substituting this x into the line's equation gives the second coordinate:
$$y = \frac{-3m^2 - 4m + 2}{3m^2 + 2}. \tag{1.5}$$
We have come a long way! Equations (1.4) and (1.5) parameterize all the rational points on the ellipse, and they in turn can be used to get all integer solutions to $2a^2 + 3b^2 = 5c^2$. We can arrive at the general solution as follows.

1.7. More General Quadratic Problems

Write m as $\frac{p}{q}$ with p, q integers, and substitute this $m = \frac{p}{q}$ into (1.4) and (1.5). After some simplification, we get for (x, y),

$$P = (x, y) = \left(\frac{3p^2 - 6pq - 2q^2}{3p^2 + 2q^2}, \frac{-3p^2 - 4pq + 2q^2}{3p^2 + 2q^2} \right).$$

Since the integers in the problem $2a^2 + 3b^2 = 5c^2$ appear only as squares, we write a solution to it using $\pm$. Our solution is, for any $m = \frac{p}{q}$,

$$\mathbf{a} = \pm(\mathbf{3p^2 - 6pq - 2q^2}),$$
$$\mathbf{b} = \pm(\mathbf{3p^2 + 4pq - 2q^2}), \qquad (1.6)$$
$$\mathbf{c} = \pm(\mathbf{3p^2 + 2q^2}).$$

As before, divide this solution (a, b, c) by its gcd (greatest common divisor) to get a primitive solution. Then any solution corresponding to $m = \frac{p}{q}$ is obtained by multiplying the primitive one by an appropriate integer whose sign may change from coordinate to coordinate.

It's now easy to find some specific solutions corresponding to choices of m. (We choose our solutions to be positive by taking absolute values.)

Example 1.7.2. $m = \frac{1}{2}$. Substituting $p = 1$ and $q = 2$ into (1.6) gives

$a = |3 - 12 - 8| = 17,$

$b = |-3 - 8 + 8| = 3,$

$c = 3 + 8 = 11,$

so $(a, b, c) = (17, 3, 11)$, which is primitive. As a check using $(a, b, c) = (17, 3, 11)$, the left-hand side of $2a^2 + 3b^2 = 5c^2$ is $578 + 27 = 605$, and the right-hand side is $5 \cdot 121 = 605$. Agreement! Even a slope as simple as $\frac{1}{2}$ gives a nontrivial result. ◇

Example 1.7.3. $m = \frac{1}{3}$. Substituting $p = 1$ and $q = 3$ into (1.6) gives

$a = |3 - 18 - 18| = 33,$

$b = |-3 - 12 + 18| = 3,$

$c = 3 + 18 = 21.$

Since $\gcd(33, 3, 21) = 3$, $(11n, n, 7n)$ ($n \in \mathbb{Z}$) represents (up to signs) all solutions for this choice of m. As a check, substitute $(11n, n, 7n)$ into

$2a^2 + 3b^2 = 5c^2$. The left side becomes $2 \cdot 121n^2 + 3n^2 = 245n^2$, and the right-hand side becomes $5 \cdot 49n^2 = 245n^2$. Agreement. ◇

We are now witnessing just the beginning of the tremendous power of our method, and it bears repeating:

> Move from a problem in number theory to a nondegenerate conic. Check to be sure a rational solution exists, and if it does, use some algebra to parameterize the conic's points. Finally, move back to the number theory setting to get specific, easily computable results.

Let's now test-drive our creation, finding an insanely unlikely solution to $2a^2 + 3b^2 = 5c^2$.

Example 1.7.4. $m = \frac{127}{283}$. Substituting $p = 127$ and $q = 283$ into (1.6) gives

$$a = |48{,}387 - 215{,}646 - 160{,}178| = 327{,}437,$$

$$b = |-48{,}387 - 143{,}764 + 160{,}178| = 31{,}973,$$

$$c = 48{,}387 + 160{,}178 = 208{,}565.$$

This (a, b, c) happens to be primitive. Let's give our result the acid test. Write (a, b, c) as

$$(327{,}437,\ 31{,}973,\ 208{,}565).$$

The left-hand side of $2a^2 + 3b^2 = 5c^2$ is

$$2 \cdot 327{,}437^2 + 3 \cdot 31{,}973^2 = 214{,}427{,}977{,}938 + 3{,}066{,}818{,}187$$
$$= 217{,}496{,}796{,}125,$$

and the right-hand side is

$$5 \cdot 208{,}565^2 = 5 \cdot 43{,}499{,}359{,}225 = 217{,}496{,}796{,}125.$$

Those two huge computed numbers agree, so $(327{,}437,\ 31{,}973,\ 208{,}565)$ is indeed a primitive solution. All solutions corresponding to the slope $\frac{127}{283}$ are therefore $(327437n, 31973n, 208565n)$ where the sign of n may vary from coordinate to coordinate. ◇

1.8. Conics

Without formula (1.6), getting this solution would truly test human capability. We could find relatively prime integer values for p and q so large that, without (1.6), the problem would bring any computer to its knees. From just the simple case of our ellipse, algebraic curves have already shown their awesome power in number theory.

Exercise 1.7.5. Find a formula for an integer solution to $3a^2 + 4b^2 = 7c^2$ and test your solution for a few rational slopes $m = \frac{p}{q}$.

Exercise 1.7.6. Repeat the above exercise for $8a^2 - 5b^2 = 3c^2$.

Exercise 1.7.7. Repeat the above exercise for $a^2 = 6b^2 + c^2$.

1.8 Conics

In this chapter the curves we're looking at all have degree at most 2. The general second-degree equation in x and y is

$$Ax^2 + Bxy + Cy^2 + Dx + Ey + F = 0, \qquad (1.7)$$

where not all of A, B, and C are zero. This equation defines the general conic, and as we know, the conics we're interested in are nondegenerate — ellipses, parabolas, and hyperbolas. We now make a basic distinction.

> Until Chapter 6, the coefficients in (1.7) are assumed to be real and, most often, rational. When real, then (1.7) and the associated general conic are called *real*. If the coefficients can be chosen to be rational, then (1.7) and the conic can be called *rational*. The conics encountered in a number theory problem are rational since they arise from dividing a number theory equation by some integer, and as a consequence (1.7) ends up being rational.

Example 1.8.1. The circle $x^2 + y^2 = \pi$ is real but not rational, while the circle $\pi x^2 + \pi y^2 = \pi$ is not only real but rational since dividing through by π gives the rational form $x^2 + y^2 = 1$. ◇

In the number theory problems we've solved up to this point, the coefficients B, D, and E have all been zero, and that means our conics are somewhat special — they're not only all "central" (that is, the center of symmetry is the origin), but the principal axes are the coordinate x- and y-axes. These are exactly the ellipses and hyperbolas plotted from equations in "standard form." Nonzero coefficients B, D, and E change all that! An xy-term always rotates the conic, and linear x- and y-terms translate it.

> The method in the last section continues to work for all quadratic problems corresponding to nondegenerate conics over the rationals. This means that even when such a homogeneous quadratic in a, b, c leads to a rational quadratic in x and y containing linear and/or mixed terms, we can still apply the approach and solve the problem.

We now illustrate our approach with two of these more general conics. The first example has tilt but no translation, and the second has both tilt and translation. In each case, simple algebra leads to the powerful solution.

Example 1.8.2. Find all integer solutions to

$$ab = c^2.$$

Dividing through by c^2 gives $\frac{a}{c} \cdot \frac{b}{c} = 1$, or $xy = 1$. This is a (rational) hyperbola whose principal axes are tilted 45° from the (x, y)-axes. A rational point on it is $P = (-1, -1)$, so there are infinitely many solutions.

To get all solutions, begin with the line of slope m through P with equation $y + 1 = m(x + 1)$. Since y is $m(x + 1) - 1$, the x-coordinates of the two points in Figure 1.6 are the solutions of $x(mx + m - 1) = 1$ — that is, the roots of $mx^2 + (m - 1)x - 1 = 0$; the quadratic formula shows that these are $x = -1$ and $x = \frac{1}{m}$. The associated y-coordinates are -1 and m, so the points are $(-1, -1)$ and $Q = (\frac{1}{m}, m)$. Write $m = \frac{p}{q}$, so that Q becomes $(\frac{q}{p}, \frac{p}{q})$. To move from the x, y world to integers a, b, and c, write Q over a common denominator: $(\frac{q^2}{pq}, \frac{p^2}{pq})$. Therefore any line through $(-1, -1)$ of rational slope $m = \frac{p}{q}$ intersects the hyperbola in just one other point,

1.8. Conics

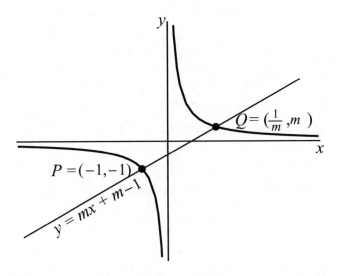

Figure 1.6. This sketch of $xy = 1$ shows the line $y = mx + m - 1$ through the rational point $(-1, -1)$. As m runs through $\mathbb{Q}$, all the points of the hyperbola in $\mathbb{Q}^2$ are covered. We'll see later on that adding points at infinity to the real plane to make it "projective," the vertical line through $(-1, -1)$ meets the hyperbola "at infinity in the projective plane."

$(x, y) = (\frac{q^2}{pq}, \frac{p^2}{pq})$. We see from Figure 1.6 that m can assume negative values, meaning that p and q can have the same or opposite signs. A little thought leads to solutions

$$\mathbf{a} = \mathbf{q}^2,$$
$$\mathbf{b} = \mathbf{p}^2,$$
$$\mathbf{c} = \mathbf{pq}$$

and

$$\mathbf{a} = -\mathbf{q}^2,$$
$$\mathbf{b} = -\mathbf{p}^2,$$
$$\mathbf{c} = \mathbf{pq}.$$

Dividing each solution (a, b, c) by $\gcd(a, b, c)$ gives a primitive solution for each m, with any other solution being an appropriate integer multiple of a primitive one. ◊

Example 1.8.3. Take $p = 3$ and $q = 5$. Then, for example,

$$a = q^2 = 25,$$
$$b = p^2 = 9,$$
$$c = pq = 15,$$

which is a primitive solution. Then $ab = c^2$ becomes $25 \cdot 9 = 15^2$, so both sides are 225. ◊

Example 1.8.4. Let $p = 7$ and $q = 11$. Then

$$a = q^2 = 121,$$
$$b = p^2 = 49,$$
$$c = pq = 77,$$

a primitive solution. In this case $ab = c^2$ is $121 \cdot 49 = 77^2$, so both sides are 5,929. ◊

Let's now showcase our method using an example with both tilt and translation.

Example 1.8.5. Find all integer solutions to

$$a^2 - 2ab + 3b^2 + 2ac - 2bc = 0.$$

Dividing through by c^2 gives $x^2 - 2xy + 3y^2 + 2x - 2y = 0$. This is a tilted ellipse going through $(0,0)$, meaning there are infinitely many solutions (and that the algebra is simpler). This ellipse and a general line through $(0,0)$ are depicted in Figure 1.7.

The algebra is especially simple because the general line through $(0,0)$ has equation $y = mx$. Substituting this into $x^2 - 2xy + 3y^2 + 2x - 2y = 0$ gives

$$x^2 - 2mx^2 + 3m^2x^2 + 2x - 2mx = 0$$

which, after factoring the left-hand side, becomes

$$x[(3m^2 - 2m + 1)x - 2(m - 1)] = 0.$$

1.8. Conics

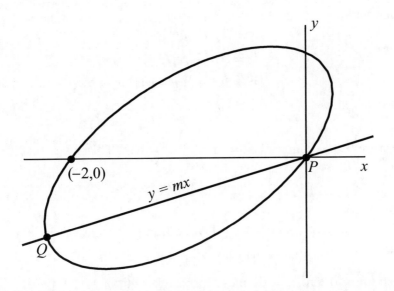

Figure 1.7. The line $y = mx$ through the origin intersects all other points Q of the rational ellipse as m runs through $\mathbb{Q} \cup \{\infty\}$. When $m = \infty$, the vertical line through the origin intersects the ellipse at $Q = (0, \frac{2}{3})$.

Set the factor in square brackets equal to zero and solve for x. This is x in the parametrization (x, y). Since y is mx, we get this for the parametrization of the ellipse:

$$(x, y) = \left(\frac{2(m-1)}{3m^2 - 2m + 1}, \frac{2m(m-1)}{3m^2 - 2m + 1} \right).$$

Writing $m = \frac{p}{q}$ and simplifying gives

$$(x, y) = \left(\frac{2(pq - q^2)}{3p^2 - 2pq + q^2}, \frac{2(p^2 - pq)}{3p^2 - 2pq + q^2} \right).$$

Corresponding to $m = \frac{p}{q}$, a solution to

$$a^2 - 2ab + 3b^2 + 2ac - 2bc = 0$$

is therefore
$$a = 2(pq - q^2),$$
$$b = 2(p^2 - pq),$$
$$c = 3p^2 - 2pq + q^2. \quad \diamond$$

Example 1.8.6. Choose $p = 1$ and $q = 5$. This gives $(a, b, c) = (-40, -8, 18)$, so
$$a^2 - 2ab + 3b^2 + 2ac - 2bc$$
becomes
$$40^2 - 2 \cdot 40 \cdot 8 + 3 \cdot 8^2 + 2 \cdot (-40) \cdot 18 - 2 \cdot (-8) \cdot 18$$
$$= 1{,}600 - 640 + 192 - 1{,}440 + 288,$$
which sums to zero, as it should. A primitive solution is $(20, 4, -9)$. $\quad \diamond$

Exercise 1.8.7. As above, find nontrivial integer solutions to
$$ab = 5c^2$$
in terms of p and q.

Exercise 1.8.8. Repeat the above exercise for
$$a^2 - 2ab + 3b^2 - 2ac + 2bc = 0.$$

1.9 Our Method Also Works in Reverse

One of the beautiful things about our method of converting a number theory problem into a geometric problem about an algebraic curve is that it works in reverse, too. That is, if you have a particular algebraic curve in mind, then as long as its defining polynomial has rational coefficients, you can see what number theory problem that curve corresponds to. (As usual, we avoid degeneracy, so a curve defined by a polynomial of degree 2 is assumed to be an ellipse, parabola, or hyperbola. In the next chapter, we will see that nondegeneracy is also important in degree 3.)

Example 1.9.1. Suppose we have a parabola such as $y = x^2$ but rotated about the origin by, say, 45°, and we'd like to find what Diophantine problem that rotated parabola corresponds to. We first need the parabola's

1.9. Our Method Also Works in Reverse

equation. To get it, observe that we can rotate around the origin any algebraic curve by θ if we feed the coordinates (x, y) of its defining polynomial $p(x, y)$ into the rotation matrix

$$R(\theta) = \begin{pmatrix} \cos\theta & \sin\theta \\ -\sin\theta & \cos\theta \end{pmatrix}.$$

In concrete terms, replace $p(x, y)$ by $p(X, Y)$ where $(X, Y) = (x, y)R(\theta)$. Then $p(X, Y)$ defines the same object as $p(x, y)$, but rotated about the origin by θ. Of course a pure magnification about the origin by a factor r doesn't change the amount θ of rotation, and such magnifying can sometimes keep both X and Y rational. That's just what we need of our new polynomial. It's straightforward to verify that when θ is 45° and $r = \sqrt{2}$, then

$$rR(\theta) = \begin{pmatrix} 1 & 1 \\ -1 & 1 \end{pmatrix}.$$

Thus $(X, Y) = (x, y)rR(\theta)$ is $(X, Y) = (x - y, x + y)$, so $p(x, y) = y - x^2$ gets replaced by

$$p(X, Y) = x + y - (x - y)^2.$$

Expanding this and setting it equal to zero gives

$$x^2 - 2xy + y^2 - x - y = 0,$$

so it's in the general form of equation (1.7) on p. 21 and we see the mixed term xy which tilts the conic. Also notice that interchanging x and y in $x^2 - 2xy + y^2 - x - y$ leaves the equation unchanged, meaning that the new parabola is symmetric with respect to the line $y = x$. This parabola is depicted in Figure 1.8.

What number theory problem does this parabola correspond to? Replacing x by $\frac{a}{c}$ and y by $\frac{b}{c}$ in $x^2 - 2xy + y^2 - x - y = 0$ and clearing denominators yields

$$a^2 - 2ab + b^2 = ac + bc.$$

Let's now solve this Diophantine equation generated by the tilted parabola. Substituting $y = mx$ into $x^2 - 2xy + y^2 - x - y = 0$ gives

$$x^2 - 2mx^2 + m^2x^2 - x - mx = 0$$

which factors into $x[(m^2 - 2m + 1)x - (m + 1)]$. Setting the second factor equal to 0 and solving for x gives $x = \frac{m+1}{m^2-2m+1}$ so, with $y = mx$, the

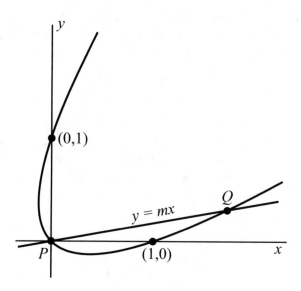

Figure 1.8. The line $y = mx$ through the origin intersects all other rational points of the parabola as m runs through $\mathbb{Q} \cup \{\infty\}$. When $m = \infty$, the vertical line through the origin intersects the parabola at $(0, 1)$.

parametrization is

$$(x, y) = \left(\frac{m+1}{m^2 - 2m + 1}, \frac{m^2 + m}{m^2 - 2m + 1} \right).$$

Now substitute $m = \frac{p}{q}$ and simplify to arrive at this solution to the number theory problem arising from the tilted parabola:

$$\mathbf{a} = \mathbf{q}^2 + \mathbf{pq},$$
$$\mathbf{b} = \mathbf{p}^2 + \mathbf{pq},$$
$$\mathbf{c} = (\mathbf{p} - \mathbf{q})^2. \quad \diamond$$

Exercise 1.9.2. In the Example 1.9.1, find a primitive solution corresponding to the slope $\frac{3}{4}$.

Exercise 1.9.3. In this same example, find a primitive solution corresponding to the slope $\frac{2}{5}$.

1.10. Surveying All Solutions

Exercise 1.9.4. Find the slope producing the solution $(a, b, c) = (6, 3, 1)$.

Exercise 1.9.5. Find the slope producing the solution $(a, b, c) = (15, 10, 1)$.

Exercise 1.9.6. In a way similar to what was used to tilt the parabola above, a rotation/expansion can also "untilt" a curve. This can be done to the Fermat curve $x^3 + y^3 = 1$ (which looks much like the right picture in Figure 1.1 on p. 4) to make the asymptote vertical. Find two matrices that rotate by $-45°$ and $135°$ and appropriately expand so one curve has rational points at $(0.5, \pm 0.5)$ and the other at $(-0.5, \pm 0.5)$.

1.10 Surveying All Solutions

Suppose a homogeneous quadratic in a, b, and c defines a nondegenerate conic, and suppose the conic has a rational point P. We've seen that any other rational point on the conic comes from selecting some line of slope $m = \frac{p}{q} \in \mathbb{Q} \cup \{\infty\}$ through P. Formulas for the two coordinates of Q in terms of p and q then lead to an integer-valued quadratic formula for each of a, b, c.

In a sense, this gives only a local view of the full answer to finding all integer solutions to the Diophantine problem. That is, in our examples so far, we've seen solutions corresponding to only one slope at a time. It's fair to ask for a broader perspective, a perspective giving an idea of what the full set of solutions looks like.

To do this, let's start by remembering that $\mathbb{Q} \cup \{\infty\}$ is countable, meaning we can list all fractions (written in lowest terms, say) using the natural numbers $\mathbb{N} = \{1, 2, 3, \dots\}$. For each member of this list — each slope — we create an integer triple (a, b, c) satisfying the Diophantine equation. Now as we've seen in examples above, the triple may have a greatest common divisor (gcd) other than ± 1, and by dividing the triple by its gcd, we arrive at a *primitive* integer triple. Now form two columns, the first being the slopes — essentially the m_n, with $n \in \mathbb{N}$ — and the second consisting of the corresponding primitive triples. Now create a third, fourth, fifth, ... column by successively multiplying each primitive triple by 2, 3, 4, ... We end up with a big $\mathbb{N} \times \mathbb{N}$ table whose entries beyond the first column are integer triples. Finally, we may further augment each triple by including triples with different signs of a, b, c, as appropriate.

This table then gives a view of the entire countable set of integer solutions to a quadratic Diophantine equation. Ignoring sign changes, the solutions (a, b, c), where each of a, b, and c is a quadratic in p and q, mean that the big table's typical entry is

$$\frac{(na, nb, nc)}{\gcd(a,b,c)}, \quad \text{where } n \in \mathbb{N}.$$

In this book, our primary goal is finding rational points on a plane rational algebraic curve coming from the given homogeneous Diophantine problem. Getting these rational points (x, y) constitutes the real meat of the problem; the transition from there to the corresponding primitive integer triples is, as we've seen in examples so far, reliably straightforward. For this reason:

> For the remainder of this book, any Diophantine problem will be posed as finding the rational points on a plane rational algebraic curve.

1.11 The Discriminant

Suppose the two-variable real quadratic equation

$$Ax^2 + Bxy + Cy^2 + Dx + Ey + F = 0$$

defines an ellipse, parabola, or hyperbola. For an easy way to tell which is which, look at the discriminant $B^2 - 4AC$. (See [Kendig 1, Chapter 9].)

- If $B^2 - 4AC > 0$, then the conic is a hyperbola.
- If $B^2 - 4AC = 0$, then the conic is a parabola.
- If $B^2 - 4AC < 0$, then the conic is an ellipse.

If we vary any or all of A, B, C, both the curve and the discriminant will likely vary in response, and if the varying is done continously and the discriminant changes sign, its value will pass through zero and we'll see the conic become a parabola or degenerate as the curves morph between ellipse and hyperbola.

The next two figures illustrate what this can look like for the parabola $x^2 - 2xy + y^2 - x - y = 0$ considered in Section 1.9. Let the coefficient of the mixed term vary — that is, let $B \in \mathbb{R}$ vary in

$$x^2 - Bxy + y^2 - x - y = 0.$$

1.11. The Discriminant

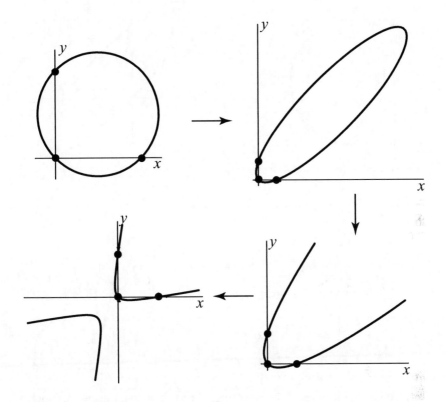

Figure 1.9. Reading clockwise from the top left sketch, this figure suggests how increasing B from zero in $x^2 - Bxy + y^2 - x - y = 0$ can lead from a circle to ellipses, then to a parabola, and on to hyperbolas.

The discriminant is $B^2 - 4$, and Figure 1.9 shows the effect of increasing B from 0, which stretches the conic. The top left sketch is of a circle with $B = 0$ and no stretching. The discriminant is -4, and because there's no xy-term, there's no tilting. As soon as B becomes nonzero, a tilt appears, as we see in the stretched-out ellipse. Encountering the sketches clockwise, we next see in the bottom right the conic corresponding to $B = 2$, which is our parabola with $B^2 - 4AC = 0$. In the next chapter we'll learn about points and the "line at infinity," and this will shed light on just how intimately ellipses, parabolas, and hyperbolas are related. For example, the parabola can be looked at as an ellipse stretched out so much that

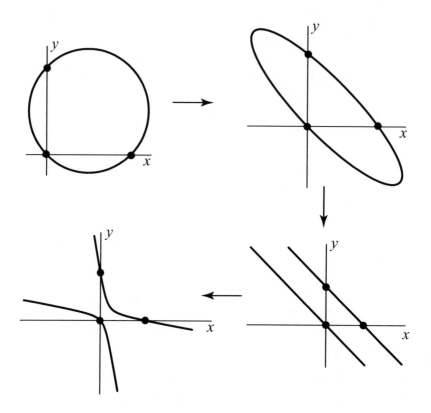

Figure 1.10. Reading clockwise from the top left sketch, this figure shows that decreasing B from zero in $x^2 - Bxy + y^2 - x - y = 0$ can in a different way lead from a circle to ellipses, to a degenerate conic of two parallel lines, and then to hyperbolas.

a point on it has actually moved onto the line at infinity. In the bottom left sketch, $B > 0$, the discriminant is positive, and the conic is a hyperbola. This value of B is large enough to have forced the ellipse over the line at infinity, and the part that crossed over appears as the branch of the hyperbola seen in the third quadrant. Notice that in all this morphing, the conic always passes through the same three points: $(0, 0)$, $(1, 0)$, and $(0, 1)$. Figure 1.10 similarly shows the effect of decreasing B from $B = 0$. When $B = -2$, the discriminant is once again zero, and the corresponding curve is degenerate — two parallel lines. Notice that again, during all this morphing, the conic continues to pass through the three points $(0, 0)$, $(1, 0)$, and $(0, 1)$.

Comment 1.11.1. Although Figures 1.9 and 1.10 are informative, continuously animating the morphing is better than a print version. This can be done using GeoGebra, Maple, or Mathematica. For information and coding for this, see Appendix C. ◇

Exercise 1.11.2. Animate the sequence of conics $Ax^2 - 2xy + Ay^2 = 1$ as A varies through nonnegative real numbers.

1.12 Finding a Rational Point

So far, it's not been a problem to find a rational point on those conics that have one. In fact, for convenience we've sometimes chosen the conic to pass through the origin. But what if we are given a more complicated equation like, say, $-9a^2 + 7b^2 + ac = 41c^2$? To use our potent method, we need to actually find a rational point on the curve. We might try different rational values for x, but after solving for y we could well find that the answer involves square roots. Such an approach could lead to a lot of hard work with only a limited chance of success. There must be a better way.

As it turns out, there is a better way. In Appendix C we present code that searches for rational points, and this works well provided the coefficients of $p(a, b, c)$ aren't too large, say between -30 and 30. This code creates a list of all integer triples (a, b, c) in the range $-10 \le a, b, c \le 10$ satisfying $p(a, b, c) = 0$.

As an interesting test case, enter $a^2 + b^2 = 3c^2$ into the program and choose k to be 1,000, say. After returning from a leisurely lunch, we see that the computer has printed out nothing new. This is as it should be, because in Section 1.5 we found two different ways of assuring us that the corresponding circle $x^2 + y^2 = 3$ has no rational points. Our computer run strongly suggests this, and we could effortlessly run a multitude of polynomials having relatively small coefficients and (likely) determine whether or not the curves contain rational points.

For larger coefficients, a method lurks in Legendre's proof of his criterion and a modern-day improvement in Legendre's approach works for really large coefficients, say 50 or 100 digits long. For integer coefficients like this, the important, much more modern paper of Cremona and Rusin (see [Cremona]) greatly improves on Legendre's approach. This method rests on Legendre's "method of infinite descent" which Legendre himself used to prove his criterion stated on p. 11. In addition to that approach, there's a concrete algorithm for finding a rational point on a rational ellipse, parabola, or hyperbola. This is the Hasse-Minkowski Theorem, which reduces the problem to one of finding a point over $\mathbb{R}$ and over "m-hour clocks" (the integers $0, \ldots, m - 1$ with addition as on an ordinary clock, but having m hours), for all positive integers m. It is beyond the scope of this book to explore these approaches, but for those with sufficient background, [Aitken] gives a good account of the first method, and J. P. Serre covers the Hasse-Minkowski Theorem in [Serre, Chapter IV].

1.13 What About Degree 1?

Our journey so far has focused on finding all integer solutions to second-degree homogeneous polynomial equations over $\mathbb{Z}$, and what we've done in this book has been a tale of success. *But what about the degree-one analog* — finding all integer solutions to degree-one homogeneous equations over $\mathbb{Z}$? The good news is that our general roadmap continues to hold. The problem is simpler, but nonetheless not trivial. There are many analogies between degree-two and degree-one problems. The basic degree-one problem takes this form:

> A degree-one equation homogeneous in the integer variables a, b, and c can be written $Aa + Bb = Cc$, with given integer coefficients A, B, C, and a, b, c being the integers we're solving for.

An example is finding all integer triples (a, b, c) satisfying $3a + 6b = 11c$. Now in the degree-two case, the first thing we did was restrict the associated curve to be nondegenerate, so that it is either an ellipse, parabola, or hyperbola, but not, say, two lines or a single point. In the degree-one case, we can write the associated line as $Ax + By = C$ (assuming c in $Aa + Bb = Cc$ isn't zero). Any real line is nondegenerate, so we needn't worry about degeneracy. Here, not both A and B are zero, so let's say $B \neq 0$. The method in the degree-two case applied to the degree-one setting tells us to find all rational points on this line. Now $Ax + By = C$ can

1.13. What About Degree 1?

be solved for y:
$$By = C - Ax \implies y = \frac{C}{B} - \frac{A}{B}x,$$

so a general point on the line is $\left(x, \frac{C}{B} - \frac{A}{B}x\right)$. This point is rational if and only if x is rational, so x is serving as a parameter for the rational points on the line. In this respect, x plays a role analogous to $m = \frac{p}{q} \in \mathbb{Q}$, so let's correspondingly write $x = \frac{p}{q} \in \mathbb{Q}$. We can now move from the geometric to the number-theoretic setting by substituting $x = \frac{p}{q}$ into
$$\left(x, \frac{C}{B} - \frac{A}{B}x\right)$$
to get
$$\left(\frac{p}{q}, \frac{C}{B} - \frac{A}{B}\frac{p}{q}\right)$$
and then multiply through to make the entries integers. Here, Bq does the job, leading to
$$a = Bp \quad \text{and} \quad b = Cq - Ap.$$

Then $Aa + Bb = Cc$ means that
$$c = \frac{Aa + Bb}{C} = \frac{ABp + BCq - ABp}{C} = Bq.$$

Assuming $B \neq 0$, we have
$$\mathbf{a = Bp,}$$
$$\mathbf{b = Cq - Ap,} \tag{1.8}$$
$$\mathbf{c = Bq.}$$

Example 1.13.1. Choose a degree-one Diophantine equation, say the one above, $3a + 6b = 11c$. Out of the infinitely many slopes $\frac{p}{q}$, randomly choose one, say $\frac{3}{5}$. Let's verify that the solution in (1.8) applied to this choice actually works. We have $A = 3$, $B = 6$, $C = 11$, $p = 3$, and $q = 5$. Then our general solution yields
$$a = 6 \cdot 3 = 18, \quad b = 11 \cdot 5 - 3 \cdot 3 = 55 - 9 = 46, \quad c = 6 \cdot 5 = 30.$$

So the triple (a, b, c) is $(18, 46, 30)$. Its gcd is 2, so $(9, 23, 15)$ is a primitive solution. Is it a solution to $3a + 6b = 11c$? Let's see:
$$3 \cdot 9 + 6 \cdot 23 = 27 + 138 = 165 = 11 \cdot 15.$$

So our primitive solution is verified in this case of $m = \frac{3}{5}$. The general solution, up to coordinatewise signs, is for this m

$$(9n, 23n, 15n), \quad n \in \mathbb{Z}. \quad \diamond$$

Exercise 1.13.2. Verify that when $B \neq 0$, the solution in (1.8) in fact satisfies the general problem. What's the story when either one of p or q is zero?

Exercise 1.13.3. Derive a companion solution to (1.8), assuming $A \neq 0$.

Exercise 1.13.4. Use appropriate search code to find an integral solution to $3a + 6b = 11c$ for which $abc \neq 0$ and $a+b+c = 3$. (See Appendix C.)

Exercise 1.13.5. Find an integral solution to $3a+6b = 11c$ so that $abc \neq 0$ and $|a| + |b| + |c|$ is as small as possible.

2

Viewing the Whole Algebraic Curve

2.1 An Ancient Indian Parable

"A group of blind people hear about a large animal called an 'elephant.' Curious, they travel to meet the great beast. One of them, feeling a leg, declares that the animal is like a tree. Another, feeling the tail, says it's actually ropelike. Another, feeling the elephant's side, decides it's basically a wall, and yet another, feeling its trunk, concludes that it is like a large snake."

In this familiar parable, each person acquires a sense of what the animal is, but that sense is only partially accurate — it's a combination of various local impressions that gets closer to reality. In this chapter we'll learn that in this respect a plane algebraic curve is like an elephant. Our aim is to get a better idea of the whole curve by exploring its various aspects.

2.2 Viewing Algebraic Curves

We begin by looking at this model problem in number theory:

> Find all solutions to $a^2 + b^2 = c^2$, with a, b, c nonzero integers.

We've always started by dividing $a^2 + b^2 = c^2$ by c^2 so that the new equation defines an algebraic curve. *But why c^2?* Why not instead divide through by a^2, or perhaps by b^2? Doing this results in different algebraic curves — hyperbolas instead of a circle. That is, dividing by a^2 gives $1 + y^2 = z^2$, where $y = \frac{b}{a}$ and $z = \frac{c}{a}$, which defines the hyperbola $z^2 - y^2 = 1$. Similarly dividing by b^2 gives $x^2 + 1 = z^2$ — the hyperbola $z^2 - x^2 = 1$, with $x = \frac{a}{b}$ and $z = \frac{c}{b}$. So now we're looking at three different algebraic curves, with each one being used to solve the identical problem in number theory. A further wrinkle: We observed in the last chapter that we can reverse the process and transition from an algebraic curve to a number theory problem. Do these three curves all yield the same problem? Look at the hyperbola $z^2 - x^2 = 1$, for example. Multiplying through by b^2 gives $c^2 - a^2 = b^2$, which of course is the same thing as $a^2 + b^2 = c^2$, but instead of finding a square equal to the sum of two other squares, the problem reads "Find a square equal to the *difference* of two other squares." It's the same thing, but stated in a less familiar form. Instead of saying that 25 is the sum of 16 and 9, it says, for example, that 16 is the difference between 25 and 9.

So far we've considered only those number theory problems whose algebraic curve is a line or an ellipse (which includes any circle), a hyperbola, or a parabola. In each case, the problem involves three integer variables a, b, and c, and the equation is homogeneous in these variables. We can easily go from the discrete (integers) to the continuous (reals) by simply allowing a, b, and c to assume real values. Doing this in place of, say, $a^2 + b^2 = c^2$, we write $x^2 + y^2 = z^2$. This is the equation of a circular cone in $\mathbb{R}^3$ whose line of symmetry is the z-axis. If we set z equal to 1, we get $x^2 + y^2 = 1$, the usual circle. Set y or x equal to 1, and we get the two hyperbolas illustrated in Figure 2.1. Geometrically, $z = 1$ defines a plane in $\mathbb{R}^3$ one unit above the (x, y)-plane, and the cone and plane intersect in that circle. Similarly, each of $x = 1$ and $y = 1$ defines a plane intersecting our cone in a hyperbola.

What are we doing here? Each of the three planes intersects the cone in an algebraic curve, and each such curve is like a sampling of the cone, the plane determining a cross section. In fact, an arbitrary plane not going through the origin of $\mathbb{R}^3$ intersects the cone in a circle, ellipse, hyperbola, or parabola — precisely the kinds of conics we've considered. Each such sampling or "photo" is like one of the blind truth seekers feeling the elephant. In a sense, the cone approximates the elephant.

2.2. Viewing Algebraic Curves

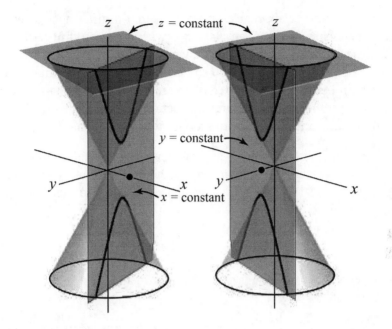

Figure 2.1. Setting z equal to a nonzero constant defines a plane parallel to the (x,y)-plane, and this plane intersects the circular cone in a circle. Similarly, setting x equal to a nonzero constant defines a plane parallel to the (y,z)-plane which intersects the cone in the hyperbola $y^2 - z^2 = $ a constant. Likewise, setting y equal to a nonzero constant gives the hyperbola $x^2 - z^2 = $ a constant.

We say "approximates" because no matter how many planes are chosen, there remain parts of the elephant that never get photographed. The planes intersecting in a hyperbola or parabola never show what happens when you ride along a branch toward "infinity." Could we perhaps fly higher and higher above the curve to get a bird's eye view, thereby getting some idea of what happens? If we try this on, say, the parabola $y = x^2$, we'll see what looks like the upper half of the y-axis, because y increases so much faster than x does when x is large. (If the unit is a millionth of an inch, then when x is one inch, y is nearly 16 miles.) Greater altitudes help even less. What about the usual plots we see in calculus texts? They're of no help at all! And looking at the curve $y = x^2$ under a microscope placed at the origin? Now we see essentially a horizontal line. Besides planes intersecting a cone, it seems that an additional approach is needed.

There is another approach, and it leads to illuminating answers. The crux of the idea is *nonuniform shrinking,* with the shrinking getting more intense the farther out we go. In fact, we want it to get so very intense that the entire plane gets pulled in to something bounded — a figure whose entirety can be drawn on a piece of paper. The unbounded figure, thus corralled, can then submit to our inspection and analysis. At that stage we will be substantially closer to seeing the entire elephant.

We can get a good idea of the new approach by seeing how it works on just the real line. We can compress the entire line down to, say, the interval $(-1, +1)$, and one way of doing this is quite simple: Map a point in the x-axis into the y-axis via

$$x \to x/\sqrt{1 + x^2} \,.$$

The graph of this mapping is shown in Figure 2.2.

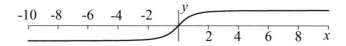

Figure 2.2. The function $x/\sqrt{1 + x^2}$ shrinks $\mathbb{R}$ down to an open interval. In this plot, all input values make up $\mathbb{R}$, or the x-axis, and all output values make up the open interval $(-1, +1)$ in the y-axis.

Notice that as we slide a unit interval in the x-axis farther and farther away from the origin, the interval gets mapped into ever smaller intervals in the y-axis. For example, under this map, $0 \to 0$ and $1 \to \frac{1}{\sqrt{2}}$, so $[0, 1]$ in the x-axis, which has length 1, maps to $[0, \frac{1}{\sqrt{2}}]$ in the y-axis, which has length about 0.707. However, move a little farther from the origin of the x-axis, and we see how quickly the compression becomes severe. For example, the unit interval $[5, 6]$ in the x-axis maps to

$$\left[\frac{5}{\sqrt{26}}, \frac{6}{\sqrt{37}}\right] \approx [.9806, .9864]$$

in the y-axis, and that has length about 0.0058 — less than a hundredth of 0.707.

2.2. Viewing Algebraic Curves

Because $x/\sqrt{1+x^2}$ approaches ± 1 as x approaches $\pm\infty$, the real line shrinks down to $(-1, +1)$. We now use this idea to shrink the entire real plane down to an open 2-disk. We do this by shrinking each line through the origin in the same way as above, and this squeezes the entire plane down to such a disk. The function doing this magic is simply the coordinatewise analog:

$$(x, y) \longrightarrow \left(\frac{x}{\sqrt{x^2 + y^2 + 1}}, \frac{y}{\sqrt{x^2 + y^2 + 1}} \right). \tag{2.1}$$

As before, line segments in the plane get more and more compressed the farther they are from the origin. To see an example of this mapping in action, consider in the plane the nine equally spaced horizontal lines

$$y = j, \quad \text{where } j = -4, \ldots, +4 \,.$$

After applying the above compression map, the image of the entire real plane with the nine lines in it looks like Figure 2.3. The increasing compression is evident and we see the parallel lines' images drawing together, approaching a common point at each end. The dashed bounding circle signifies that the image of the real plane under our mapping is an open disk.

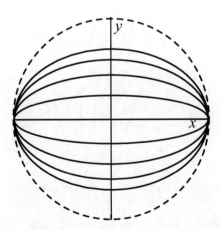

Figure 2.3. The image in the disk of nine horizontal lines under our compression map.

42   Chapter 2. Viewing the Whole Algebraic Curve

If we add in the disk's boundary points, then the "ends" of any two lines intersect in two points. We know two parallel lines in the real plane don't intersect at all, and now, by adding the boundary points, their images intersect in two points — too much of a good thing! To make them intersect in one point the way lines should, the secret is to identify opposite points on the closed disk. Parallel lines don't intersect in the finite plane, but now, with identified opposite boundary points added, they intersect in exactly one point. We call any such identified boundary point a *point at infinity*, and all the points at infinity form a line called the line at infinity. Let's make these ideas official:

Definition 2.2.1. In a disk, we call two diametrically opposite boundary points, when identified to a point, a *point at infinity*. The set of all points at infinity of the disk is called the *line at infinity*. The disk with opposite points identified is *the real projective plane*, or *the disk model of the real projective plane*. We denote the real projective plane by $\mathbb{P}^2(\mathbb{R})$.   ◊

The line at infinity is in fact a line, and any two distinct lines (even if one is the line at infinity) intersect in exactly one point. Figure 2.4 illustrates the gist of this.

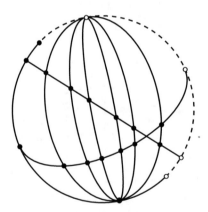

Figure 2.4. In the projective plane, every line under our mapping turns out to be a "semi-ellipse" — that is, one of the parts connecting the ends of the ellipse's major axis. When the ellipse is a circle, then that circle, with opposite points identified, is the line at infinity. In this figure, choosing any two different lines we've drawn shows the two lines intersecting in just one point, even when one of the lines is the line at infinity.

2.2. Viewing Algebraic Curves

We have seen what happens to parallel lines under our compression map. We now give a few examples of other curves in $\mathbb{R}^2$ that have been compressed via the mapping (2.1) to make them lie in a disk. Our first example helps clarify the story for parabolas.

Example 2.2.2. In the real plane, the branches of any parabola $y = x^2 - c$, where c is a constant, become more and more vertical the farther we travel away from the origin. Although they're not vertical lines, the branch slopes increase without bound and as Figure 2.5 suggests, for any constant c, this increase is fast enough so that in the disk, the images of its branches bend toward each other and approach a common point on the disk's boundary — that is, a point at infinity. To actually carry out plotting the image in the disk, parameterize the parabola in $\mathbb{R}^2$ by

$$t \to (t, t^2 - c), \quad t \in \mathbb{R},$$

and then divide both x (the first coordinate) and y (the second coordinate) by $\sqrt{x^2 + y^2 + 1}$ as we do in (2.1). A parametrization of the disk image is

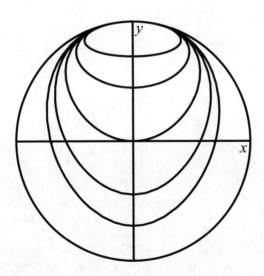

Figure 2.5. Images in the projective disk of the parabola $y = x^2$ and its translates up and down by $\frac{1}{2}$ and 1. In each case the added point at infinity completes the image to a closed loop.

therefore

$$t \to \left(\frac{t}{\sqrt{t^2 + (t^2 - c)^2 + 1}}, \frac{t^2 - c}{\sqrt{t^2 + (t^2 - c)^2 + 1}} \right), \quad t \in \mathbb{R}.$$

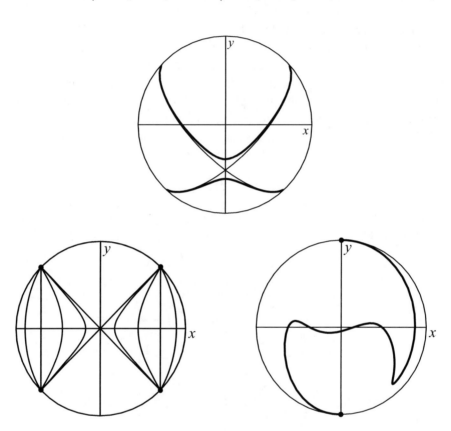

Figure 2.6. In the top picture, the parabola $y = x^2 - 1$ in Figure 2.5 has been pushed across the line at infinity, becoming a hyperbola translated downward from its standard position. Its asymptotes are translated too, each becoming half an ellipse in the projective disk. The bottom left picture depicts the images in the projective disk of the four hyperbolas $x^2 - y^2 = c^2$ with $c^2 = .175, .5, 1, 2$. At the bottom right we see the image of an odd-degree polynomial in the projective plane. More generally, the image in $\mathbb{P}^2(\mathbb{R})$ of any polynomial $y = p(x)$ with real coefficients is a topological loop — intuitively, a rubber band in which no two of its points touch.

2.2. Viewing Algebraic Curves

Figure 2.5 shows the images for c equal to 0, $\pm\frac{1}{2}$, and ± 1. ◇

Example 2.2.3. The top drawing in Figure 2.6 shows the parabola $y = x^2 - 1$ in Figure 2.5 after being "pushed across the line at infinity." (See Comment 2.2.10 at the end of this section.) Since opposite points are identified, as the pushing occurs, pairs of points at the top of the disk also appear at the bottom, with the result that the parabola gets split into upper and lower parts — we call them branches — and the figure now gets called a hyperbola. Note that since opposite points are identified, the image is still a closed loop. ◇

Example 2.2.4. Shrinking $\mathbb{R}^2$ under the map

$$t \to \left(\frac{t}{\sqrt{t^2 + (t^2 - c)^2 + 1}}, \frac{t^2 - c}{\sqrt{t^2 + (t^2 - c)^2 + 1}} \right), \quad t \in \mathbb{R},$$

can lead to surprises. The bottom left sketch of Figure 2.6 depicts the images of the four square hyperbolas $x^2 - y^2 = c^2$ for $c^2 = .175, .5, 1, 2$. The nature of the shrinking causes the image of $x^2 - y^2 = 1$ to appear as two vertical line segments. For any $c > 1$ the branches of $x^2 - y^2 = c^2$ bow toward the disk's boundary. The two line segments of slope ± 1 are the common asymptotes to the hyperbolas for all c^2. In the disk model, when $c^2 = 1$ or 2, they don't seem to act like asymptotes — that is, "tangent at infinity." Angles are not preserved under the shrinking map, and at infinity the notion of distance itself breaks down, so the metric notion of asymptotic may not look the way we expect. However, any hyperbola in the disk $\mathbb{P}^2(\mathbb{R})$ forms a closed loop. ◇

Example 2.2.5. Figure 2.7 shows a much fancier example of an algebraic curve, drawn in both the usual (x, y)-plane and in the disk. The dashed lines in the plane and in the disk are asymptotes. Notice in the disk how the algebraic curve's "ends" meet at the line at infinity, and the points of intersection there connect the parts that seem separate in the plane so that the disk curve is all one connected piece. ◇

> Since any points at infinity of a curve get included in the projective disk, we will often draw figures in $\mathbb{P}^2(\mathbb{R})$ as well as in $\mathbb{R}^2$.

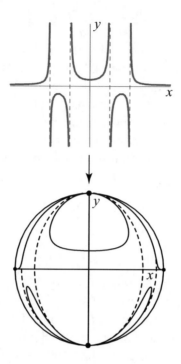

Figure 2.7. The plane curve $y = 1/\bigl((x^2 - 1)(x^2 - 4)\bigr)$ is mapped into the disk. The locus is unchanged when we multiply this equation through by the denominator, so the curve is algebraic.

Manually drawing curves in the projective disk can lead to inaccuracies, so it's generally better — and often very enlightening — to use software such as GeoGebra, Maple, or Mathematica to get results that are more reliable. The parametric mapping

$$(x, y) \longrightarrow \left(\frac{x}{\sqrt{x^2 + y^2 + 1}}, \frac{y}{\sqrt{x^2 + y^2 + 1}} \right)$$

from $\mathbb{R}^2$ into the unit open disk leads to a parametric mapping for the projective image of a curve defined by $x = x(t)$, $y = y(t)$. Just substitute $x(t)$ for x and $y(t)$ for y in the above mapping and let $t \in \mathbb{R}$ vary over an appropriate range. For completeness, the projective image should be supplemented with the bounding unit circle as well as the axes. See Appendix C for code producing the projective image of a parameterized plane curve.

2.2. Viewing Algebraic Curves

Exercise 2.2.6. After consulting Appendix C if necessary, use a computer to draw pictures in the projective disk of $y = x^2 + c$ for $c = 0, \pm 1$. Do the same for some horizontal translates of $y = x^2$.

Exercise 2.2.7. In a way similar to how the circular functions cos and sin parameterize the circle (or "square ellipse") $x^2 + y^2 = 1$, the hyperbolic functions cosh and sinh can be used to parameterize the square hyperbola $x^2 - y^2 = 1$. Use this idea to get code producing the bottom left picture in Figure 2.6.

Exercise 2.2.8. Create code that plots the projective image of a polynomial of degree 5, making it look similar to the lower-right drawing in Figure 2.6. Also, experiment with a few polynomials of degree 6.

Comment 2.2.9. Notice in Figure 2.6's top drawing that the two asymptotes are quite curved. If the hyperbola were down even lower, the point where the asymptotes cross would also be down farther, the two asymptotes would be even more curved, and their two points of intersection with the line at infinity would get closer to each other on the circle. In fact, as the intersection point nears the disk boundary, the asymptotes get ever closer to the circular boundary. ◇

Comment 2.2.10. The phrase "push across the line at infinity" is meant to intuitively capture the process depicted in Figure 2.8. In (a) we see an oriented circle in the projective plane, and in (b), that circle has been pushed so it just touches the line at infinity. We may think of the circle in (b) as being tangent to the line at infinity. Because opposite points are identified in the projective model, point P where the circle touches the line at infinity also appears antipodally and is again labeled P. Sketch (c) depicts the situation after pushing the circle still farther. Sketch (d) depicts the circle after having been pushed entirely across the line at infinity. Notice how identifying opposite points causes the orientation given to the original circle to switch, so an orientation given to a little circle doesn't stay consistent as it's moved around in the projective plane. In other words, the projective plane is "nonorientable." The same thing happens as a little oriented circle drives all the way around a roadway's centerline painted on a Möbius strip. The Möbius strip is therefore also nonorientable. ◇

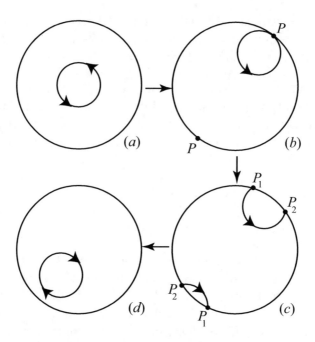

Figure 2.8. Reading clockwise, we see the oriented circle shown in the upper-left picture being pushed across the projective plane's line at infinity, ending up as a circle with reversed orientation.

2.3 Doing Math at Infinity!

We can use the power of algebra to accomplish a remarkable feat: We can transform a picture of an algebraic curve in, say, the (x, y)-plane, to one where we're sitting at a point at infinity and can survey that curve from that vantage point. This makes it possible to do some actual mathematics at infinity. For example, the two ends of a curve might meet at a point at infinity, and that curve might cross the line at infinity transversally. (A line through a point P of a curve crosses the curve transversally at P if it's not tangent to the curve there.) However, it might instead be tangent to the line at infinity, or it may even happen that the curve has an inflection point on the line at infinity. We can see the geometry and carry out computations at these points the same way as at any finite point before the transformation. The secret to doing this is to first "homogenize" the defining polynomial $p(x, y)$.

2.3. Doing Math at Infinity! 49

Definition 2.3.1. If $p(x, y)$ is a polynomial of degree n, then we *homogenize* p by multiplying each of its terms of degree less than n by some power of another variable (say z) to make the term have degree n. The resulting polynomial $p(x, y, z)$ is then homogeneous of degree n (as defined in Definition 1.7.1 on p. 16). We say $p(x, y, z)$ is the *homogenization of* $p(x, y)$; $p(x, y, z)$ is then homogeneous of degree n. In $p(x, y, z)$, setting any one of x, y, or z equal to a nonzero constant such as 1 *dehomogenizes* $p(x, y, z)$ with respect to that variable, and the new polynomial is then a *dehomogenization* of $p(x, y, z)$ with respect to that variable. ◇

Example 2.3.2. If $p(x, y) = x^2 - y^2 - x + y - 1$, then its homogenization is $x^2 - y^2 - xz + yz - z^2$. We can then dehomogenize with respect to x, y, or z by setting that variable equal to 1. Doing this to y, for example, produces the curve $x^2 - 1 - xz + z - z^2 = 0$ lying in the (x, z)-plane. ◇

Example 2.3.3. Suppose an algebraic curve in the (x, y)-plane consists of the three vertical lines $x + 1 = 0$, $x = 0$, and $x - 1 = 0$. The product equation $(x+1)x(x-1) = 0$ defines the union of these three parallel lines, because any point on any of these lines makes some factor zero, meaning the whole product is zero. Since any point not on any of these lines makes each factor nonzero, the product is nonzero. It's easy to check that we can homogenize $(x + 1)x(x - 1)$ by homogenizing each of its factors. Doing this gives $(x + z)x(x - z)$. Dehomogenizing by setting y equal to 1 (there is no y, so we're not doing anything) gives $(x + z)x(x - z)$, defining the union of the lines $x + z = 0$, $x = 0$, and $x - z = 0$. These all cross at one point at the "end" of the y-axis. The top half of Figure 2.9 shows the lines in the projective disk before and after relocating to infinity.

The dehomogenization $(x + z)x(x - z)$ tells us how to draw the new figure, and the geometry tells us how to label its coordinates. The important fact here is that dehomogenizing with respect to a variable defines an inversion of that variable. In this example, for instance, dehomogenizing with respect to y flips the y-axis via the inversion $y \to y^{-1}$, so that the parts of the original y-axis around the origin and around the point at infinity are interchanged in the new picture. Specifically, the three principal coordinate axes in the disk model are the x-, y-, and z-axes, and in the original view the z-axis is the line at infinity. The coordinates at the "end" of the y-axis are the y-axis and the z-axis and these map to the new origin. Similarly, the original x- and y-axes get mapped to infinity. These mappings are depicted in the bottom part of Figure 2.9. ◇

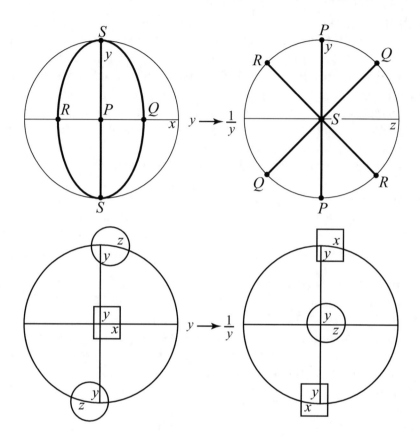

Figure 2.9. Top: Three heavily drawn lines in the disk before and after dehomogenizing to send the point S at infinity to the origin. Bottom: The (y, z)-coordinates at infinity (circled) map to the new origin; the original (x, y)-coordinates (boxed) map to infinity. The y-axis stays vertical. Note axis labels in the top right picture.

Here are some further examples:

Example 2.3.4. $y^2 = x^3 - x$ defines a cubic curve. Its plot is depicted on the left in Figure 2.10. Homogenizing this equation gives $y^2z = x^3 - xz^2$, and dehomogenizing at $y = 1$ yields $z = x^3 - xz^2$, plotted on the right. The sketches in Figure 2.11 depict the projective disk views of Figure 2.10. ◊

2.3. Doing Math at Infinity!

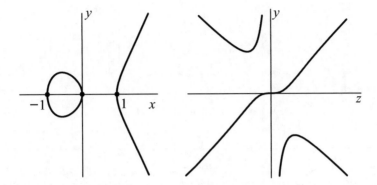

Figure 2.10. The drawing on the left depicts the cubic $y^2 = x^3 - x$; the upper and lower parts of its branch meet at the "end" of the y-axis. Homogenizing and then dehomogenizing by setting $y = 1$ allows us to determine the nature of that point at infinity. In the right sketch, the point at infinity where the branches meet appears at the origin. In these new coordinates, it can be shown that the origin is a point of inflection. The x-axis in the left drawing becomes the new line at infinity in the right drawing, and because the oval crosses it, the oval becomes the two branches in the second and fourth quadrants.

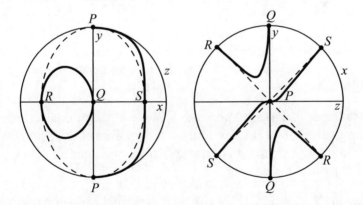

Figure 2.11. In this projective view of the affine sketches in Figure 2.10, the x-, y- and z-axes correspond to each other in the right and left pictures. Ditto for the points P, Q, R, S. The y-axis is vertical in both pictures but gets mapped into itself via $y \to y^{-1}$.

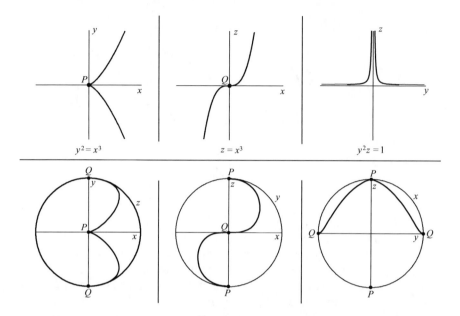

Figure 2.12. The top row depicts the dehomogenizations at $z = 1$, $y = 1$, and $x = 1$ of the homogenized equation $y^2z = x^3$ of a cusp curve and below, the corresponding images in the projective disk.

Example 2.3.5. For an educational picture, look at Figure 2.12. The top row shows three dehomogenizations of $y^2z = x^3$ at $z = 1$, $y = 1$, and $x = 1$, and just below each of these is the corresponding image in the projective disk. ◇

Exercise 2.3.6. Let $p(x, y)$ and $q(x, y)$ be two polynomials. Prove that the homogenization of their product is the product of their homogenizations. (Therefore it doesn't matter whether we homogenize before or after multiplying out the product.)

Exercise 2.3.7. Let $p(x_1, \ldots, x_N)$ be a polynomial of degree n. Prove p is homogeneous of degree n if and only if $p(tx_1, \ldots, tx_N) = t^n p(x_1, \ldots, x_N)$.

Exercise 2.3.8. A set in $\mathbb{R}^N$ is said to be *homogeneous* if and only if it is the union of lines through the origin. Prove that the zero set of a polynomial $q(x_1, \ldots, x_N)$ is homogeneous if and only if q is homogeneous.

2.4 A More Symmetric Model of $\mathbb{P}^2(\mathbb{R})$

The disk model of the real projective plane can be looked at physically as a rubber disk lying on a table top. With its boundary held fixed on that table top, pump in air underneath the disk to make the disk inflate into a "northern hemisphere," and now attach to this another hemisphere — a "southern hemisphere" — to make a full sphere. Next, identify each pair of the sphere's antipodal points; we have in this way created another model of the real projective plane — one that's more symmetric. This is not only more beautiful but has some advantages. First, it allows more flexibility in what we call the line at infinity. Rather than the sphere's equator being *the* line at infinity, we can choose any great circle on the sphere to play this role. Either hemisphere of what's left after deleting any great circle is then the associated affine part. (For us, "affine curve" means the curve lies in a real plane.) Importantly, this sphere model provides a natural way to give the whole real projective plane a topology. An open set in a basis of the topology is formed this way: Draw a tiny circle on the sphere, and let the small cap defined by this circle — minus the circle itself — be an open set. The points of this set are identified with diametrically opposite points, so the little open set appears as a pair of caps on the sphere. Relative to this topology, the boundary of any affine part is the line at infinity. The points at infinity of any affine curve get included in the curve's topological closure.

3

Entering the World of Elliptic Curves

3.1 Curves of Degree 3

In our journey so far, number theory problems have been of degree 1 or 2, with the associated rational curve being a line, ellipse, parabola, or hyperbola. We now look at degree 3. Our general method of turning a number theory problem into one of finding rational points on a rational curve is like striking a gold-bearing vein. So to any third-degree problem we associate a rational cubic curve. Although geometry will guide us as before, the new landscape will look quite different. For example, a quadratic number theory problem either has no solutions or a countably infinite number of them, while third-degree problems can have a positive, finite number of solutions as well. Here are some examples showing what can happen:

Example 3.1.1. Find all integer solutions to $3a^3 + 4b^3 + 5c^3 = 0$. Assuming $c \neq 0$, divide through by c^3, getting $3x^3 + 4y^3 + 5 = 0$. It turns out that this equation defines an affine curve which avoids passing through even one rational point in $\mathbb{R}^2$. So unless $c = 0$, there are no integer solutions. Even when $c = 0$, then $-\frac{3}{4} = \left(\frac{b}{a}\right)^3$ as long as $a \neq 0$. However, the cube root of $-\frac{3}{4}$ isn't rational, so the only integer solution to $3a^3 + 4b^3 + 5c^3 = 0$

is $a = b = c = 0$. For $c \neq 0$, this rational curve's plot in $\mathbb{R}^2$ is depicted in Figure 3.1. In $\mathbb{Q}^2$, the plot is empty. ◇

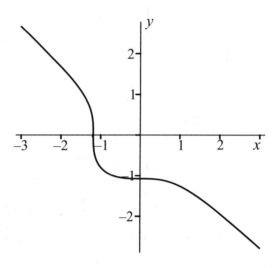

Figure 3.1. The rational cubic $3x^3 + 4y^3 + 5 = 0$ succeeds in avoiding every rational point in the real plane.

Example 3.1.2. The number theory problem $a^3 + b^3 = c^3$ has an associated affine curve $x^3 + y^3 = 1$, another example of a Fermat curve. This looks similar to the right sketch of Figure 1.1 on p. 4. There are exactly two rational points on it: $(1, 0)$ and $(0, 1)$, a consequence of Fermat's Last Theorem. ◇

Example 3.1.3. For the problem $a^3 - ac^2 = b^2c$, divide the equation by c^3, with $x = \frac{a}{c}$ and $y = \frac{b}{c}$ leading to the affine curve $y^2 = x^3 - x$. We met this curve at the end of the last chapter, and its plot in the (x, y)-plane is depicted in the left sketch in Figure 2.10 on p. 51. The points $(-1, 0)$, $(0, 0)$, and $(1, 0)$ where the curve crosses the x-axis turn out to be the only rational points of this affine curve. ◇

Example 3.1.4. The problem $a^3 - a^2c = b^2c + bc^2$ leads to $y^2 + y = x^3 - x^2$ which defines the cubic curve depicted in Figure 3.2. In the rational plane, this affine curve consists of just the four points $(0, 0)$, $(1, 0)$, $(0, -1)$, and $(1, -1)$. ◇

3.2. What Is an Elliptic Curve?

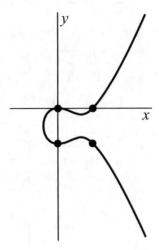

Figure 3.2. The only rational points in the affine cubic defined by $y^2 + y = x^3 - x^2$ are the four corners of a square.

Example 3.1.5. The curve $y^2 = x^3 - x + 1$, depicted in Figure 3.18 on p. 79, looks like many other curves with only finitely many rational points, but this one happens to have infinitely many of them. ◇

3.2 What Is an Elliptic Curve?

In solving a homogeneous number theory problem of the third degree, we associate a cubic curve. Compared to quadratics, there's a much larger variety of cubics — the algebraic form of the general third-degree equation in two variables is

$$a_0x^3 + a_1x^2y + a_2xy^2 + a_3y^3 + a_4x^2 + a_5xy + a_6y^2 + a_7x + a_8y + a_9 = 0. \quad (3.1)$$

That's a lot of coefficients, and this is reflected in the large array of possible shapes. Figure 3.3 shows just six possibilities.

So far, we've assumed our curves are nondegenerate, and we now similarly toss out cubics exhibiting degenerate behavior. An example of degeneracy is the union of three lines. Another is the union of an ellipse and a line, and in a moment we'll revisit this example. Excluding degenerates

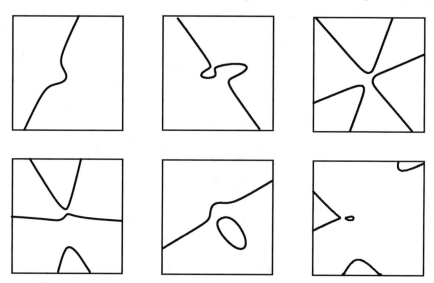

Figure 3.3. Compared to nondegenerate conics, nondegenerate cubics come in a much wider variety of shapes.

leads to this definition of *elliptic curve*:

Definition 3.2.1. An *elliptic curve* $C \subset \mathbb{P}^2(\mathbb{R})$ over $\mathbb{Q}$ is a smooth curve containing a rational point and defined by an irreducible cubic polynomial $p(x, y) \in \mathbb{Q}[x, y]$. (Here, $\mathbb{Q}[x, y]$ stands for the set of all polynomials in x and y with rational coefficients, and "irreducible" means $p(x, y)$ can't be factored into lower-degree terms in $\mathbb{Q}[x, y]$.) ◇

Comment 3.2.2. In non-number-theoretic settings, we sometimes consider elliptic curves whose defining polynomial has coefficients in $\mathbb{R}$ rather than $\mathbb{Q}$, and we say the curve is *defined over* $\mathbb{R}$. In that case, the curve must contain a real point. More generally, if the curve is defined over a field K, the required point must have coordinates in K. ◇

Definition 3.2.1 uses three ideas which we now explain. First, $C \subset \mathbb{P}^2(\mathbb{R})$ means that if $p(x, y) \in \mathbb{R}[x, y]$ defines a curve in $\mathbb{R}^2$, then it defines a curve in $\mathbb{P}^2(\mathbb{R})$ by taking the topological closure of the curve's disk image relative to the topology on $\mathbb{P}^2(\mathbb{R})$ described on p. 53. This amounts to adding in the curve's points at infinity. In the next section we define and justify requiring irreducibility, and in the section after that, we explain what "smooth" means.

3.3 Why Irreducible?

We begin with a general fact.

> In the real plane, if $f(x,y) = 0$ defines the set S and $g(x,y) = 0$ defines T, then $f(x,y) \cdot g(x,y) = 0$ defines $S \cup T$.

Proof. On the one hand if P is in $S \cup T$, then it's in at least one of S and T, so at least one of f, g is zero at P; hence their product is zero there. Conversely, if fg is zero at P, then at least one of f, g is zero there, meaning P is in at least one of S and T, so P is in $S \cup T$. Note that this argument works equally well over $\mathbb{Q}$, over $\mathbb{C}$, and also in n dimensions.

Now look at the two curves in Figure 3.4.

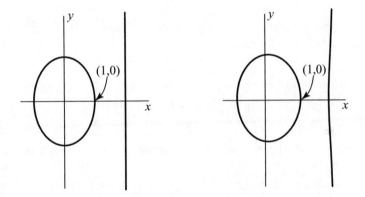

Figure 3.4. A reducible and an irreducible cubic.

The curve on the left is the union of an ellipse and a line. Its defining polynomial is reducible — the product of a polynomial defining the ellipse, and another defining the line. The curve on the right has equation $(2x^2 + y^2 - 1)(x - 2) + 1 = 0$, which happens to be irreducible and defines a card-carrying elliptic curve. Notice that the curve contains the rational point $(1,0)$. These curves look so much alike that, after just a quick glance, one might think they're identical. In truth, the facts about rational points on them are worlds apart. The ellipse in the left picture might or might not contain a rational point, so its set of rational points is either empty or there's "$\mathbb{Q}$'s worth of them." Intuitively, "$\mathbb{Q}$'s worth of them" means that a decal with the members of the set $\mathbb{Q} \cup \{\infty\}$ printed on it fits perfectly over the entirety of the ellipse. A similar thing holds for the vertical line, because if the x-coordinate is irrational, then the set of rational points is empty, while if it is rational, the line contains a dense copy of $\mathbb{Q}$. These facts about the ellipse and the line are independent of each other. We can slightly vary the ellipse and line to get any one of four scenarios: empty-empty, dense-empty, empty-dense, or dense-dense. In contrast, the set of rational points of the elliptic curve in the right picture turns out to be "dense," meaning that in both the oval and the branch, one finds no gaps of positive length in either set, no matter how intensely you magnify about any point. It can never happen for any elliptic curve that, say, the oval has no rational points and the branch has a dense assortment of them. The reason we can get mixed behavior in the ellipse-plus-line is that each part is independently defined by a polynomial. Such a reducible polynomial can lead to mixed behavior, while in the irreducible case, that doesn't happen. If the polynomial isn't a mixture (that is, not a product) of different lower-degree polynomials, then the number-theoretic behavior won't be mixed, either. Therefore:

> For the rest of this book, we'll always assume a polynomial $p(x, y)$ is irreducible unless stated otherwise.

Irreducibility excludes equations of the form $(ax + by)(cx + dy) = 0$, since the solution set of this equation is a degenerate conic, in this case the union of two lines rather than an ellipse, parabola, or hyperbola. There is in fact a rational point on this conic — the origin. But rational points in these lines are not in one-to-one correspondence with lines of rational slope through the origin via intersection as we've seen with ellipses, hyperbolas, and parabolas. This shows yet another reason for always assuming that $p(x, y)$ is irreducible.

Exercise 3.3.1. True or false? A polynomial $p(x, y) \in \mathbb{R}[x, y]$ is irreducible if and only if its homogenization is. Prove your answer.

3.4 Smoothness

It turns out that a good definition of smoothness relies on some of the wisdom mathematicians have gained over the years: To get the most consistent and beautiful results about algebraic curves, it's best to work in

3.4. Smoothness

the complex setting rather than the real. A simple example demonstrates some of the reasoning behind this.

Example 3.4.1. Consider two circles intersecting in two points. When you pull the circles away from each other, the intersection points move, at some stage getting very close to each other. When the circles become tangent, the two points have coalesced into one double point. But if we continue to pull the circles farther away from each other, those two points disappear. Where have they gone? Although restricting our vision to $\mathbb{R}^2$ offers no obvious clue, we can algebraically solve for the two common solutions to the circles. When they intersect in two separate points, the algebra produces two distinct solutions, and when they're tangent, we get two identical solutions. These solutions nicely track what we see with our eyes. The power of algebra is such that we can still solve for the common solutions even when the circles no longer intersect! Algebra becomes a wise guide, handing us two complex solutions rather than leaving us empty-handed the way looking in $\mathbb{R}^2$ did. These complex solutions represent two points in $\mathbb{C}^2$ rather than in $\mathbb{R}^2$, so by working in the complex setting, we've eliminated an exception and gained consistency. ◇

Comment 3.4.2. Just as the equation for either of the above circles can be regarded as defining a solution set in $\mathbb{C}^2$, the same is true for any curve defined by a polynomial $p(x, y) \in \mathbb{R}[x, y]$. It is in this spirit that we make the following definition. ◇

Definition 3.4.3. Let $p(x, y)$ be irreducible in $\mathbb{R}[x, y]$. The curve $p(x, y) = 0$ in $\mathbb{C}^2$ is *smooth at* (x_0, y_0) if and only if $p(x_0, y_0) = 0$ and at least one of $p_x(x_0, y_0)$ and $p_y(x_0, y_0)$ is defined and nonzero. If the curve is smooth at each of its points in $\mathbb{C}^2$, then it is *smooth* in $\mathbb{C}^2$ as well as in $\mathbb{R}^2$. The term *nonsingular* may be substituted for "smooth" throughout this definition. ◇

Comment 3.4.4. We will meet the complex projective plane $\mathbb{P}^2(\mathbb{C})$ later, in Section 6.2. If a curve in $\mathbb{P}^2(\mathbb{C})$ is smooth at each of its points in each affine part, then we say the projective curve is *smooth*. ◇

There's a companion to "smoothness":

Definition 3.4.5. In Definition 3.4.3, if the curve $p(x, y) = 0$ in $\mathbb{C}^2$ isn't smooth at (x_0, y_0), then (x_0, y_0) is called a *singularity* of the curve and the curve is said to be *singular* there. ◇

Therefore a point Q is singular if both partials p_x and p_y are zero there. Some intuition lying behind this "partials" definition of smoothness is supplied shortly after the statement of the Implicit Function Theorem in Appendix A.

Comment 3.4.6. After all we've said, you may still be wondering "What's so bad about saying a curve in the real plane is smooth provided at least one of the partials p_x and p_y is nonzero at each of its real points?" The problem is that being smooth at all the real points doesn't prevent a singularity from popping up at some nonreal point. In the case of all the elliptic curves we've been discussing, any of those real curves actually lies on the full curve in the complex setting, and that full curve should look like a torus. We'll see how basic that is in Chapter 6 (most fundamental properties depend on it) and a singularity would ruin that, preventing the full curve from being a torus.   ◇

The next few examples will help make the above definitions more meaningful.

Example 3.4.7. We can apply the fact in the box on p. 59 to $f(x, y) = x$ and $g(x, y) = y$. Since $x = 0$ defines the complex y-axis and $y = 0$ defines the complex x-axis, $p = xy$ defines their union. Now $p_x = y$ and $p_y = x$. Both partials are zero at $(0, 0)$, so the origin, where the complex lines cross, is a singular point of C defined by $p(x, y) = xy$. From the definition, it's easy to check that C is smooth at every other point. All of this agrees with what we see in $\mathbb{R}^2$.   ◇

Example 3.4.8. A curve $C \subset \mathbb{R}^2$ consisting of two parallel lines can be defined, say, by $x^2 = 1$. Although C is smooth in $\mathbb{C}^2$, it isn't smooth even in $\mathbb{P}^2(\mathbb{R})$ since the point at infinity is on both lines, and it is singular. Algebraically, here's the argument: Homogenizing the polynomial $x^2 - 1$ gives $x^2 - z^2$. Since y doesn't appear in this, the dehomogenization at $y = 1$ remains $x^2 - z^2 = (x + z)(x - z)$, which defines two lines that cross at the new origin. The partials there are both zero.   ◇

Example 3.4.9. Two intersecting lines can be thought of as a self-intersecting curve, but a self-intersecting curve consisting of just one part is more typical. One example is an "alpha curve" such as $y^2 = x^3 + x^2$, whose plot is depicted in Figure 3.5. We see that the curve crosses itself at $(0, 0)$. The partials of $p(x, y) = y^2 - x^3 - x^2$ are $p_x = -3x^2 - 2x$ and

3.4. Smoothness

$p_y = 2y$. Both of these partials are zero at the origin, which means that the curve has a singularity there. This cubic isn't smooth, so it's not an elliptic curve. ◇

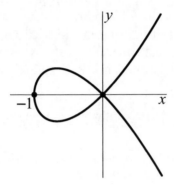

Figure 3.5. An "alpha curve" $y^2 = x^3 + x^2$.

Example 3.4.10. Another type of singular point is the cusp. Here, instead of the curve self-intersecting as in Figure 3.5, the curve doubles back on itself, as in Figure 3.6. This is another example of a cubic curve that's not elliptic — its defining polynomial $p(x, y) = y^2 - x^3$ has partials p_x and p_y that vanish at the origin. ◇

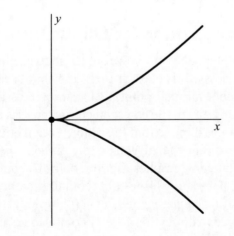

Figure 3.6. A basic cusp curve $y^2 = x^3$.

Comment 3.4.11. A real curve may look smooth, but not actually be smooth according to our definition. A good example is the "valentine curve" illustrated on p. 175. This has a singularity at the origin, but we don't see any evidence for it in the real plane. See Appendix A for an explanation. The geometric and algebraic worlds reflect each other much better in the complex setting than in the real. One reason we humans spend so much time on real curves is because viewing them is easy, while for most people, viewing curves in the complex setting is far more difficult. ◇

Example 3.4.12. For a more subtle example, consider the curve defined by
$$p(x,y) = (x^2 + y^2 - 1)(x^2 + 4) = 0.$$
Its plot in the real plane is an ordinary circle centered at the origin — nothing more, nothing less, even in $\mathbb{P}^2(\mathbb{R})$. Yet the curve is not smooth! It has four singular points at $(\pm 2i, \pm\sqrt{5}) \in \mathbb{C}^2$. ◇

Exercise 3.4.13. We assume our conics are nondegenerate. Can we now just as well say that we assume our projective conics are smooth?

Exercise 3.4.14. Why isn't the projective curve defined by $y = x^3$ smooth?

Exercise 3.4.15. Find a curve that intersects the line at infinity in two points, singular at one and nonsingular at the other.

3.5 Weierstrass Forms for Elliptic Curves

The general form of a cubic in two variables shown in equation (3.1) on p. 57 can lead to messy formulas. It turns out that because the curve is assumed to contain a rational point, (3.1) can be massaged into shorter forms without losing any of the essential characteristics of elliptic curves. This massaging can be done so that the rational point ends up at the "end" of the y-axis and so that the curve is triply tangent there to the line at infinity. ("Triply tangent" implies that the point is a point of inflection.) All this is possible if and only if the cubic contains a rational point, and in that case the transition can be carried out via "birational transformations." The algebra involved in doing this is fairly complicated and we omit it; it's explained well in [Silverman, pp. 22–24].

There are two standard shorter forms. Incredibly, one of them has only two parameters instead of ten! It's called the *Weierstrass short form*,

3.5. Weierstrass Forms for Elliptic Curves

or simply the *short form*, and is the one we'll most often use in this book. This Weierstrass short form is

$$y^2 = x^3 + ax + b. \tag{3.2}$$

There is also a *Weierstrass long form*:

$$y^2 + a_1 xy + a_3 y = x^3 + a_2 x^2 + a_4 x + a_6. \tag{3.3}$$

Although the long form has more coefficients, these coefficients usually require fewer digits and can be dramatically smaller than in the short form. Actually, long-form coefficients are never greater in size than short form coefficients. In developing lists or databases of elliptic curves, the numerators and denominators of coefficients can become horrendously large, so the long form can be more efficient. These forms are named after the German mathematician Karl Weierstrass (1815–1897) who made fundamental discoveries about "elliptic functions"; these functions can parameterize elliptic curves in much the same way that sines and cosines can parameterize ellipses, or hyperbolic sines and hyperbolic cosines can parameterize hyperbolas.

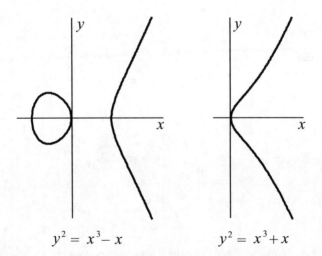

Figure 3.7. Short-form elliptic curves typically have one of these two general shapes.

The short form has an important advantage: In it, the only occurrence of y is as y^2, so the graph is symmetric about the x-axis, a fact we'll continually exploit. The real graph of an elliptic curve given in short form tends to look like one of the two pictures shown in Figure 3.7. We have already seen an example whose plot has just one real component like the picture on the right, namely $y^2 = x^3 + 2x + 3$. The curve on the left has two components, with one piece looking much like the right-hand curve, but it also has that oval piece on the left. The unbounded part is called the curve's *identity component*. (We'll see the reason for this name in the next section.) Any elliptic curve given in short form either comes with an oval or it doesn't. It has an oval exactly when $x^3 + ax + b = 0$ has three distinct real roots, because then there are three points of intersection with the x-axis, and it turns out that two of them are always connected by an oval. Since y occurs only as y^2, both the identity component and the oval are symmetric about the x-axis. The short form makes it easy to distinguish between the two types of curve.

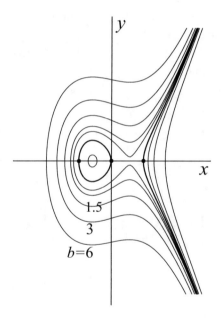

Figure 3.8. Nine examples of elliptic curves in Weierstrass short form. Only two of these curves have an oval.

Figure 3.8 shows a more detailed version of Figure 3.7. Each of the nine curves in this figure has either one or three real roots. All these curves have short-form equations $y^2 = x^3 - x + b$, with b ranging from 6 to

3.5. Weierstrass Forms for Elliptic Curves

-3. We can identify any curve in Figure 3.8 by looking at its intersection with the x-axis. Going from left to right, the first five curves correspond to b equal to $6, 3, 1.5, 0.7, 0.4$, and each of these curves intersects the x-axis in just one point. The next curve, drawn more heavily, corresponds to $b = 0$ and is one we'll often see. As we know, it has an oval and intersects the x-axis in the three points $x = -1, 0, +1$. The next curve corresponds to $b = -0.35$ and it too has an oval, which is smaller, and the curve's rightmost intersection is the next one to the right of the dot appearing at $x = 1$. The last two curves correspond to b equal to -1 and -3 and each has a single intersection with the x-axis.

By the way, putting an elliptic curve into short form is a really major step because it casts a general elliptic curve into one with symmetry about the x-axis, giving the curve an easy-to-recognize form. Looking at just a few general elliptic curves as in Figure 3.3 on p. 58 lets us appreciate what putting elliptic curves into short form does for us. All six curves there are elliptic, but all lack that nice symmetry. The coefficients of these curves were chosen so that the frequency of their shapes decreases as you go left to right and top to bottom. Roughly speaking, in choosing general curves of degree 3 using random coefficients, the shape looking most like the upper left is encountered about a third of the time, while that in the lower right is seen less than 1% of the time.

Exercise 3.5.1. Verify that for any curve in short form, its point in $\mathbb{P}^2(\mathbb{R})$ at the "end" of the y-axis is a point of inflection.

Exercise 3.5.2. There are important examples of elliptic curves that aren't symmetric about the x-axis, but rather about some other horizontal line. Find the long Weierstrass form for an elliptic curve symmetric about the line $y = y_0$.

Exercise 3.5.3. Starting with the short form $y^2 = x^3 - x$, what is the geometric effect of keeping the constant term $b = 0$ fixed and varying a, the coefficient of x? Visually check your answer using the freeware GeoGebra discussed in Appendix C.

Exercise 3.5.4. Is the curve in $\mathbb{P}^2(\mathbb{R})$ defined by
$$(y-1)^2 = (x-1)^3 - (x-1)$$
elliptic? What is its line of symmetry? Find three (finite) rational points.

3.6 The Discriminant Again

The discriminant of the quadratic $ax^2 + bx + c$ is

$$b^2 - 4ac,$$

assuming $a \neq 0$. The two solutions to $ax^2 + bx + c = 0$ are

$$x = \frac{-b \pm \sqrt{b^2 - 4ac}}{2a},$$

and their difference

$$\frac{-b + \sqrt{b^2 - 4ac}}{2a} - \frac{-b - \sqrt{b^2 - 4ac}}{2a}$$

simplifies to

$$\frac{\sqrt{b^2 - 4ac}}{a}.$$

This difference of the quadratic's roots is zero precisely when the discriminant $b^2 - 4ac$ is zero. It turns out that the concept of discriminant generalizes to any polynomial $p(x)$ of degree n with real or complex coefficients, and the above root difference is at the core of the notion. Here's the idea. If the polynomial's n roots are $r_1, r_2, \ldots, r_n$ (these may not be distinct), then the discriminant is essentially the squared product of their differences,

$$\prod_{j>i}(r_j - r_i)^2.$$

The differences are squared to ensure that the discriminant is a polynomial in the coefficients of $p(x)$, in much the same way that $\sqrt{b^2 - 4ac}$ is squared to obtain the polynomial $b^2 - 4ac$.

The discriminant turns out to be an important tool in studying elliptic curves. For example, the discriminant of a cubic can determine in one fell swoop whether or not the cubic curve is smooth and, if smooth, whether the curve consists of just a branch or has an oval in addition. The discriminant is quite simple for the right-hand side of the short form, $x^3 + ax + b$:

Definition 3.6.1. The discriminant of $x^3 + ax + b$ is

$$\Delta = -(4a^3 + 27b^2). \tag{3.4}$$

In the context of elliptic curves, this is also taken to be the discriminant of $y^2 = x^3 + ax + b$. ◇

3.6. The Discriminant Again

As we'll see in a moment, just being in short form doesn't ensure the equation defines an elliptic curve. Here's a central fact:

> A cubic $y^2 = x^3 + ax + b$ defines an elliptic curve if and only if $\Delta = -(4a^3 + 27b^2)$ is not zero.

As we've said, the discriminant is zero exactly when at least two roots of the polynomial equation $x^3 + ax + b = 0$ are the same. Let's explore what this means. One example of when $x^3 + ax + b = 0$ has three distinct roots is $x^3 - r^2 x = 0$ — that is,
$$(x+r)x(x-r) = 0$$
for any nonzero $r \in \mathbb{R}$. The roots are 0 and $\pm r$. In the short form, $a = -r^2$ and $b = 0$, so $\Delta = 4r^6$ which is nonzero, as expected. But we can separately change these three roots in $(x+r)x(x-r)$ to make any two or all three coalesce. Taking a cue from the discriminant's form $\Delta = -(4a^3 + 27b^2)$, one obvious choice is $a = b = 0$, giving the curve $y^2 = x^3$, the cusp curve depicted in Figure 3.6. In this case, all three roots have coalesced at $x = 0$. Another case is $a = -3$ and $b = 2$, and this gives an alpha curve, $y^2 = x^3 - 3x + 2$, depicted in Figure 3.9. The roots are -2, 1, and 1.

Yet another case is $a = -3$ and $b = -2$. The plot is depicted in Figure 3.10 and represents a third way of coalescing roots. The roots of $x^3 - 3x - 2$ are $-1, -1$, and 2. This curve's plot consists of a branch through $x = 2$ opening to the right, plus an isolated double point at $x = -1$. It's easy to check that for $p(x, y) = y^2 - (x^3 - 3x - 2)$, the partials p_x and p_y are zero at $(-1, 0)$, so $(-1, 0)$ is a singular point of this cubic.

There are many more ways the discriminant can vanish. In fact, it's easily checked that for any real r, the cubic $y^2 = x^3 - 3r^2 x + 2r^3$ has discriminant zero. More generally, within the (a, b)-parameter space $\mathbb{R}^2$, the coefficients a, b for which the discriminant is zero form a cusp curve, depicted in Figure 3.11.

Comment 3.6.2. The probability that a randomly chosen point in the (a, b)-plane lies exactly on the cusp in Figure 3.11 is zero. That's because once you pick a y-value y_0, there's only one corresponding x-value x_0 for which (x_0, y_0) is on the curve. The number of x-values in $\mathbb{R}$ is infinite, so

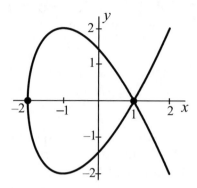

Figure 3.9. Simply being in Weierstrass short form does not ensure that the curve is elliptic. This figure shows an example since $y^2 = x^3 - 3x + 2$ is in short form, but it doesn't define an elliptic curve since the curve has a singularity at $x = 1$.

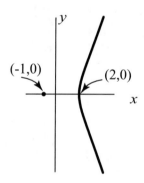

Figure 3.10. This curve defined by $y^2 = x^3 - 3x - 2$ is another example showing that just being in short form doesn't guarantee the curve is elliptic. Here, $(x, y) = (-1, 0)$ is a double point.

the probability of randomly hitting x_o is zero, meaning that a randomly chosen point P in the (a, b)-plane has a 100% chance of not lying on the cusp curve. In this sense, then, the short-form curve corresponding to P has a 100% chance of being elliptic.   ◇

3.6. The Discriminant Again

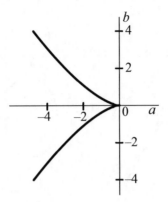

Figure 3.11. This cusp curve in (a, b)-space parameterized by $\{a = -3r^2, b = 2r^3, r \in \mathbb{R}\}$, consists of those (a, b) making the discriminant of $y^2 = x^3 + ax + b$ zero. Eliminating the parameter r gives us the curve's equation $27b^2 = -4a^3$.

As an interesting observation, consider $y^2 = x^3 - x + b$. The discriminant of this elliptic curve is $\Delta = (4 - 27b^2)$, and this is zero exactly when $b^2 = \frac{4}{27}$ — that is, when $|b| = \frac{2}{3\sqrt{3}} \approx 0.3849$. The discriminant is positive when $b^2 < \frac{4}{27}$ and negative when $b^2 > \frac{4}{27}$. If Δ is zero, the corresponding curve turns out to be alpha-shaped, its crosspoint making it nonsmooth and therefore not elliptic. Figure 3.12 shows one way an elliptic curve can morph from having two components to just one as the discriminant changes sign. For the elliptic curve $y^2 = x^3 - x + b$ on the left, $|b|$ was chosen to make $\Delta = (4 - 27b^2)$ negative, while for the curve on the right, $|b|$ was decreased a bit to make Δ positive.

Figure 3.12 suggests one instance of this general fact:

> An elliptic curve $y^2 = x^3 + ax + b$ in $\mathbb{R}^2$ consists of a single branch if and only if its discriminant $\Delta = -(4a^3 + 27b^2)$ is negative. The curve consists of a branch plus an oval if and only if its discriminant is positive.

Exercise 3.6.3. Consulting Appendix C if necessary, create an animated version of Figure 3.12.

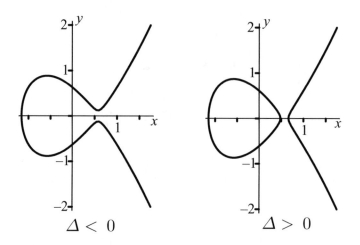

Figure 3.12. This depicts one way an elliptic curve can change its number of components as Δ changes sign.

Exercise 3.6.4. Is there some polynomial $p(x, y) \in \mathbb{R}[x, y]$ whose zero set in $\mathbb{R}^2$ has an isolated point and is smooth there? (A point P is isolated in a curve provided all other points on the curve are at least some distance $\epsilon > 0$ away from P.)

Exercise 3.6.5. Later in this book we'll meet $y^2 = -x^3 - ax + b$, the reflection about the y-axis of the elliptic curve $y^2 = x^3 + ax + b$. Its branch always opens to the left rather than always to the right. What is this reflected curve's discriminant?

3.7 Our Method Still Works

The basic method we've used to solve quadratic Diophantine equations is to divide a homogeneous equation in integer variables by the square of one of those variables to get an algebraic curve where we can apply geometry to help solve the problem. This general method also works for problems of degree 3, where we now divide by the cube of one of the integer variables. Just as we looked only at quadratic number theory problems whose associated curve is a nondegenerate conic over $\mathbb{Q}$ and contains a rational point to actually have solutions, we now look at only third-degree

3.7. Our Method Still Works

number theory problems whose associated curve is defined over $\mathbb{Q}$, is nondegenerate, and contains a rational point. That means the coefficients of the defining polynomial can be chosen to be rational and that the curve is smooth. As we said earlier, third-degree problems can be far more subtle than quadratic ones. For one thing, the idea of finding one rational point P on a conic and then finding all other solutions via rational-slope lines through P no longer works. That led to an "all or nothing" result for quadratics, while elliptic curves may have a finite positive number of points or, as we'll see, different degrees of "infinitely many" points. In place of rational-slope lines through P used for quadratics, we'll use ideas like tangent lines and symmetry to get a geometric algorithm that leads to one or several successions of rational points on the cubic. Starting with a (guaranteed) rational point, the algorithm sometimes produces new rational points forever. Other times, depending on the curve and/or the starting point, it can simply cycle around a finite set of rational points.

The algorithm works most clearly when the curve's equation is in Weierstrass short form. We present our algorithm using a succession of pictures, since that conveys the spirit of the method better than using only words. To begin, take a look at Figure 3.13, a plot of an elliptic equation in Weierstrass short form. A line through two points P_1 and P_2 on the curve is guaranteed to intersect the curve in a third point because if on one hand P_1 and P_2 lie on a nonvertical line $L: y = mx + q$, then substituting $mx + q$ into the short form produces a cubic in x. That cubic has three roots, so there are three (not necessarily distinct) corresponding points (x, y), two of them being P_1 and P_2. If P_1 and P_2 are rational, then the third point is also rational. The argument is essentially the same as in the quadratic curve case for a line through a single point P. (We consider in a moment the other case, when P_1 and P_2 lie on a vertical line.)

We then reflect the third point about the x-axis. Why do we do that? If we didn't, those three points on the line would get stuck in an endless three-point cycle and not produce any new rational points since any two points in our line determine the remaining point.

> We will often refer to the algorithm just described as the "connect-the-dots" algorithm or method.

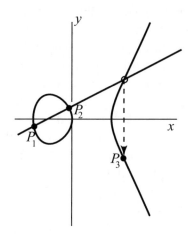

Figure 3.13. The line through rational points P_1 and P_2 intersects the curve in a third rational point whose reflection about the x-axis is the new rational point P_3.

A particularly nice feature of this connect-the-dots algorithm is that we can start from just one rational point P. To do this, think of a single point as a double point by letting a point P' on the curve approach P. The secant line through $P' \neq P$ approaches the tangent line to the smooth curve at P as P' approaches P. Figure 3.14 shows how $P_1 = P_2$ generates P_3.

This is a very powerful observation, since it means that a single rational point on the cubic can often start the algorithm.

Example 3.7.1. Let's look again at $y^2 = x^3 - x$, which we met earlier and for convenience is plotted again in Figure 3.15. The only rational points on this affine curve happen to be the three on the x-axis. The horizontal line through any two of these yields the third since reflection is trivial on the x-axis. ◇

Example 3.7.2. A really remarkable thing about our connect-the-dots algorithm is that it works even for vertical tangent lines. In Example 3.7.1 the tangent line through each of the three rational points is vertical. To appreciate what happens in such a case, take a look at Figure 3.16, the disk model of our curve $y^2 = x^3 - x$. In this figure, the tangent line through each rational point of $y^2 = x^3 - x$ of course remains vertical.

3.7. Our Method Still Works

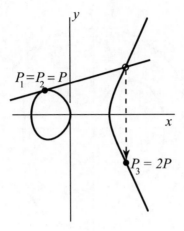

Figure 3.14. The connect-the-dots algorithm works starting with a single point.

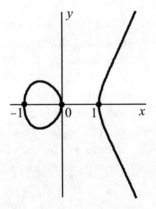

Figure 3.15. The well-behaved elliptic curve $y^2 = (x+1)x(x-1)$.

The one through $(-1, 0)$ is dashed, and it meets the curve at infinity. The reflection about the x-axis gives the same point since antipodal points at infinity are identified.

76 Chapter 3. Entering the World of Elliptic Curves

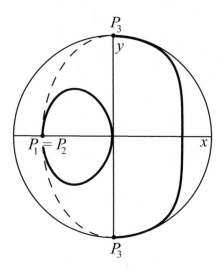

Figure 3.16. All tangent lines through rational points on the curve $y^2 = x^3 - x$ are vertical.

At the point at infinity, what point does our connect-the-dots algorithm produce next? Repositioning ourselves at this point at infinity (after homogenizing and then dehomogenizing at $y = 1$) visually suggests that this point is a triple point — a point of inflection — which we noted earlier at the beginning of Section 3.5. Figure 3.17 shows the situation. The curve's equation after repositioning is $z = x^3 - xz^2$, and implicitly differentiating twice with respect to x shows that the second derivative of z with respect to x at $(0,0)$ is zero, as the curve's sketch suggests. This means that the algorithm endlessly produces the point at infinity. ◊

Comment 3.7.3. In Example 3.7.2 above, the curve $y^2 = x^3 - x$ was homogenized and dehomogenized with respect to y to yield $z = x^3 - xz^2$. Differentiating $z = x^3 - xz^2$ twice with respect to x showed that the second derivative of z with respect to x at $(0,0)$ is zero, making $(0,0)$ an inflection point. How does differentiating $z = x^3 - xz^2$ twice with respect to x fit in with "coordinates done right" as discussed in conjunction with Figure 2.9 on p. 50? In these new coordinates, the equation of the curve $z = x^3 - xz^2$ becomes $y = z^3 - zy^2$. This defines the same curve, only in different coordinates. So differentiating $z = x^3 - xz^2$ twice with respect

3.7. Our Method Still Works

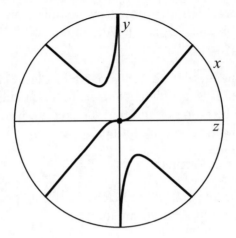

Figure 3.17. Here, the curve $y^2 = x^3 - x$ has been homogenized and dehomogenized with respect to y to yield $z = x^3 - xz^2$. The curve's inflection point is now the dot at the origin.

to x simply becomes differentiating $y = z^3 - zy^2$ twice with respect to z — essentially just a change in notation. ◇

Exercise 3.7.4. Working from a short-form picture of an elliptic curve, show how to reverse the sequence of new points generated by the connect-the-dots algorithm. That is, if our algorithm generates the sequence of points
$$\ldots, P_i, P_{i+1}, P_{i+2}, \ldots,$$
show geometrically how to generate
$$\ldots, P_{i+2}, P_{i+1}, P_i, \ldots.$$

Hand-calculating successive points on an elliptic curve using our algorithm can be time consuming and prone to error. An antidote to this is having a computer do it. Appendix C contains one approach to constructing such code.

Comment 3.7.5. A major reason for studying elliptic curves is that they can be so useful in solving cubic number theory problems. Such a curve's equation comes from dividing a homogeneous cubic number theory equation in integers a, b, and c through by the cube of one of these integers to get a two-variable cubic equation with rational coefficients. Because of this:

> We assume until Chapter 6 that all elliptic curves are rational (that is, coefficients of the defining cubic can be chosen to be rational) and that the points we consider on such an elliptic curve are rational.

We can even go a step further:

> We may assume that all coefficients in an equation defining a (rational) elliptic curve are integers.

To see why, we may assume that the elliptic curve's equation is in short form: $y^2 = x^3 + ax + b$, with $a, b \in \mathbb{Q}$. Now replace x and y by $x = \frac{X}{\gamma^2}$ and $y = \frac{Y}{\gamma^3}$, where $\gamma \in \mathbb{Z}$. (This stretches the curve horizontally by γ^2 and vertically by γ^3.) Then $y^2 = x^3 + ax + b$ becomes $\frac{Y^2}{\gamma^6} = \frac{X^3}{\gamma^6} + a\frac{X}{\gamma^2} + b$. Multiplying through by γ^6 then gives

$$Y^2 = X^3 + a\gamma^4 X + b\gamma^6.$$

By choosing the integer γ to be a common multiple of the denominators of a and b, we can ensure that the coefficients in the displayed equation are all integers. This same idea also works for an equation in Weierstrass long form. ◇

3.8 Examples of Connecting the Dots

For some elliptic curves our connect-the-dots method yields only finitely many rational points. For others, it will spit out brand new rational points forever. Here are some quick examples.

Example 3.8.1. In each of the first four examples at the beginning of this chapter, the elliptic curve has only finitely many rational points. ◇

Example 3.8.2. The curve $y^2 = x^3 - x + 1$ has infinitely many rational points, and it turns out that the single point $P_1 = P_2 = (1, -1)$ on the curve can initiate the connect-the-dots algorithm. Figure 3.18 shows how

3.8. Examples of Connecting the Dots

this point generates P_3. It is remarkable that the points and their reflections produced by the algorithm account for every single rational point on this elliptic curve. Of all the infinitely many rational points on this curve, whichever one we choose is eventually produced this way. So after many iterations, the computed rational point might be light-years away from the origin, and after further iterations, the rational point could be less than a micron from the origin. ◇

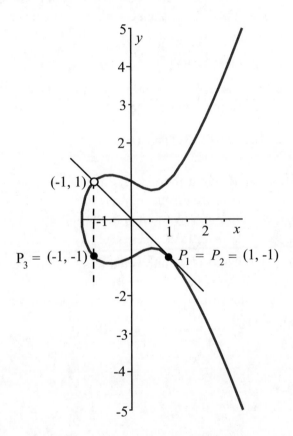

Figure 3.18. This depicts the first step of the connect-the-dots algorithm applied to $y^2 = x^3 - x + 1$, starting at the double point $P = P_1 = P_2 = (1, -1)$. The tangent line through this point intersects the curve in the second quadrant, and this intersection, reflected about the x-axis, is $2P = (-1, -1)$. One can check that $3P = (0, 1)$ and $4P = (3, 5)$.

Exercise 3.8.3. Using the connect-the-dots algorithm, show that in Figure 3.18, $3P = (0, 1)$ and $4P = (3, 5)$.

Whenever P is a rational point of a curve $y^2 = x^3 + ax + b$, so is its reflection about the x-axis, since reflecting just changes the sign of the y-coordinate. This gives a natural way to index the rational points of the above curve $y^2 = x^3 - x + 1$ — successive point pairs can be indexed by $0, \pm 1, \pm 2$, and so on, thus filling out a copy of the integers $\mathbb{Z}$.

3.9 Mordell's Theorem

One of the most consequential results in studying elliptic curves is Mordell's Theorem. Here is what it says:

Theorem 3.9.1 (Mordell's Theorem).

> On any elliptic curve there exists a finite set of rational points so that starting from this set, the connect-the-dots routine can produce any rational point on the curve.

Louis Mordell (1888–1972) established this famous theorem in 1922; a proof appears in [Husemöller, Chapter 6]. The theorem lays the foundation for one of the most prominent aspects of elliptic curves. We explore this in the next chapter.

4
Every Elliptic Curve Is a Group!

We begin with an apology. Most of the curves in the last chapter were chosen to be affine and in Weierstrass short or long form, but we never put a spotlight on one of the most consequential things about any of these elliptic curves: its point at infinity. Importantly, *that point must be rational.* Why is that? At the beginning of Section 3.5 on p. 64, we noted that by using birational transformations, any elliptic curve can be massaged into long as well as short Weierstrass form, implying that the guaranteed rational point on any elliptic curve winds up at the "end" of the y-axis — that is, on the line at infinity. Its very name "birational transformation" suggests it maps rational points into rational points, and this suggestion is right — the transformed elliptic curve's point at infinity is indeed rational. And by opening the curtains and letting in sunshine, we meet an extraordinary aspect of the rational points on an elliptic curve — *they form an abelian group.*

4.1 Finite Groups Within an Elliptic Curve

We begin with a definition:

Definition 4.1.1. A *commutative group* (also called an *abelian group*) is a set G supplied with a binary operation $+ : G \times G \to G$ for which these axioms are satisfied:
- Commutativity: $a + b = b + a$ for all a, b in G.
- Associatively: $(a + b) + c = a + (b + c)$ for all a, b, c in G.
- Existence of identity: There is a unique element 0 in G satisfying $a + 0 = 0 + a = a$ for all a in G.
- Existence of inverse: For each a in G there is a unique element $-a$ satisfying $a + (-a) = (-a) + a = 0$. ◇

Example 4.1.2. Examples of abelian groups are:
- $\mathbb{Z}$, the set of all integers under addition.
- $\mathbb{Z}_m$, the set of integers modulo an integer m.
- $\mathbb{Q}$, the set of all rational numbers under addition.
- $\mathbb{R}$, the set of all real numbers under addition.
- $\mathbb{C}$, the set of all complex numbers under addition.
- $\mathbb{Q}^+$, the set of all positive rational numbers under multiplication, with identity 1 and $1/a$ as inverse of a.
- $\mathbb{R}^+$, the set of all positive real numbers under multiplication.
- $\mathbb{R}^n$, a real n-dimensional vector space under vector addition. Algebraically, vector addition means adding *coordinatewise* — that is,

$$(a_1, \ldots, a_n) + (b_1, \ldots, b_n) = (a_1 + b_1, \ldots, a_n + b_n).$$

- No one, two, or three quadrants of $\mathbb{R}^2$ ever forms a group, since at least one of the group conditions is violated under vector addition. ◇

We now look at some concrete examples of commutative groups residing in elliptic curves.

Example 4.1.3. We start with $y^2 = x^3 - x$, one of our favorite elliptic curves. Its plot is depicted in Figure 3.15 on p. 75, and the only rational points on this real affine curve are, as we've said, its three intersections with the x-axis. Figure 3.16 on p. 76 does show its point at infinity, but now look at Figure 4.1. This is a redrawing of Figure 3.16 and it suggests something astonishing lies beneath the surface.

4.1. Finite Groups Within an Elliptic Curve

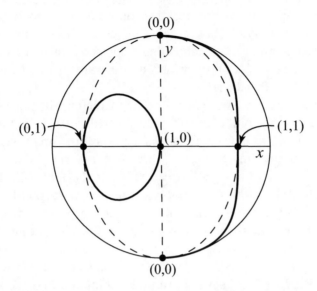

Figure 4.1. The four rational points are relabeled to show that they are elements of the group $\mathbb{Z}_2 \oplus \mathbb{Z}_2$.

Here is the remarkable fact:

> Our connect-the-dots algorithm is also an *addition algorithm*.

This means that not only does our algorithm find new rational points (which in general saves a lot of guesswork), the algorithm actually says that its output point is the sum of its two input points (including when both input points are the same).

This leads to another remarkable fact:

> Our addition algorithm turns the set of rational points of any elliptic curve into a commutative (or abelian) group.

Commutativity is clear since when two points determine a line, the order in which they are considered is irrelevant. Any elliptic curve in long or short form has a branch whose "ends" meet at a point at infinity, and that point serves as the group's identity. Each element must have an inverse, in this case its negative. Finally, any group must be associative.

We give an easy argument for negatives on p. 86. The much longer and not-so-easy argument for associativity is given in [Silverman, pp. 19–21]. An alternative approach to establishing associativity is to let a computer algebra package such as Mathematica do the grunt work. One resource explaining this is [Fuji-Oike].

Let's now try to make sense of the new labelings in Figure 4.1. For example, why is the point at infinity now labeled $(0,0)$? In Figure 3.17 on p. 77, we saw that this point is actually a triple point — an inflection point — so after choosing any two of the three equal points, the tangent line between them passes through the third, which is also that point at infinity. Then reflecting about the x-axis puts us at the antipodal point on the line at infinity, but since these antipodal points are identified, reflection does nothing, and our algorithm merely produces the same point at infinity that it started with. If our algorithm is actually an addition algorithm on points, then we can write $P + P = P$, with P now considered an element of a group; $P + P = P$ implies $P = 0$ and the point at infinity serves as the zero element of our abelian group of rational points.

Next question: Why are the points labeled as ordered pairs? Well, look at any of the curve's three points on the x-axis, say $(0, 1)$, and let's add that point to itself, meaning we're taking $(0, 1)$ to be a double point. The line through that double point is tangent to the curve and is vertical. From Figure 4.1, we see that the algorithm gives $(0, 0)$. That is, $(0, 1) + (0, 1) = (0, 0)$. Similarly, $(1, 0) + (1, 0) = (0, 0)$ and $(1, 1) + (1, 1) = (0, 0)$. So each element in the group is its own additive inverse. Furthermore, notice what happens when we add different points on the x-axis to each other: $(0, 1) + (1, 0) = (1, 1)$, $(0, 1) + (1, 1) = (1, 0)$, and $(1, 1) + (1, 0) = (0, 1)$. These results exactly describe addition in the 4-element group $\mathbb{Z}_2 \oplus \mathbb{Z}_2$. So the addition algorithm applied to the four rational points on the elliptic curve $y^2 = x^3 - x$ turns these four points into the additive group $\mathbb{Z}_2 \oplus \mathbb{Z}_2$.
◇

Examples teach and illuminate, so let's continue.

Example 4.1.4. Another example of an elliptic curve with only a finite number of rational points is $y^2 = x^3 + 4x$. Its affine and projective plots are depicted in the next figure. In the disk model, the four points are labeled as the elements of $\mathbb{Z}_4$, where 0 is the point at infinity and 1 is chosen to be $(2, 4)$. We can see this cyclic group structure by successively adding 1 to $0, 1, 2, 3$. The projective line through points 0 and 1 intersects the curve

4.1. Finite Groups Within an Elliptic Curve

at point 3, which, when reflected about the x-axis, gives 1, so $0 + 1 = 1$. Another line through 1 is the tangent line to the curve there, and this line can be shown to intersect the curve at point 2, so $1 + 1 = 2$. That same line through 1 and 2 intersects the curve in the other double point at 1, and reflecting lands us at point 3, meaning that $2 + 1 = 3$. Finally, the (dashed) arc connecting 1 and 3 intersects the curve at 0, so $3 + 1 = 0$. Notice that the vertical tangent through point 2 intersects 0, whose reflection is itself, so $2 + 2 = 0$. And just as $1 + 1 = 2$, analogously, $3 + 3 = 2$. All these different sums exactly describe the cyclic group $\mathbb{Z}_4$. ◇

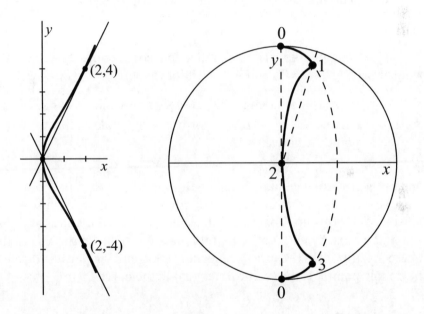

Figure 4.2. The curve on the left is the affine plot of $y^2 = x^3 + 4x$ and shows the rational points, $(0,0)$, $(2,4)$, and $(2,-4)$. There's a fourth at infinity. We see that fourth point in the disk model on the right, where the four rational points are relabeled as the elements $\{0, 1, 2, 3\}$ of a cyclic group $\mathbb{Z}_4$ of four elements. (We also say the group has *order* 4.)

86   Chapter 4. Every Elliptic Curve Is a Group!

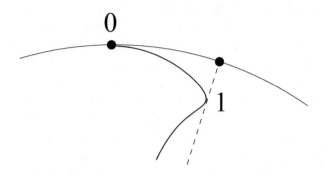

Figure 4.3. This magnified part of the previous figure's right drawing shows that the line between points 1 and 2 is actually tangent to the curve.

Notice that in this example, reflection corresponds to taking the negative: $-1 = 3, -2 = 2$, and $-3 = 1$. More generally:

> Reflecting about the x-axis any point P of an elliptic curve $y^2 = x^3 + ax + b$ corresponds to taking the point's negative, $-P$.

One can justify this by taking the line through P and its reflection. This line is vertical and intersects the elliptic curve at its point at infinity — the 0-element of its group. That is, 0 is the sum of P and something, and that something must be $-P$. Notice that if P happens to be on the x-axis, the argument still works because its reflection is P again, and the line through the double point P is the (vertical) tangent to the curve there.

Example 4.1.5. The elliptic curve defined by $y^2 = x^3 + 1$ happens to have exactly six rational points. Figure 4.4 depicts an affine view on the left, where five of these rational points are visible. The sketch on the right shows all six points in the curve's disk view.   ◇

Example 4.1.6. The cubic $y^2 + y = x^3 - x^2$ isn't in Weierstrass short form, but the curve's branches still meet at the "ends" of the y-axis, which is the 0-element of the group of five rational points. The four other group elements sit at the corners of a square with (x, y)-coordinates $(0, 0), (1, 0), (0, -1)$, and $(1, -1)$. In using our addition algorithm, reflection is always

4.1. Finite Groups Within an Elliptic Curve

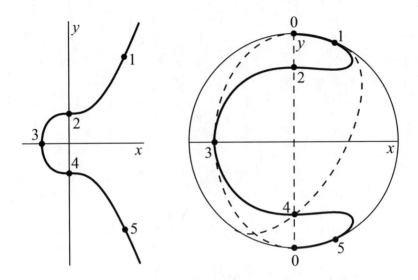

Figure 4.4. This depicts the affine and projective views of the curve defined by $y^2 = x^3 + 1$. The rational points are labeled as group elements of $\mathbb{Z}_6$, with the projective view showing the 0-element at infinity. Elements 2 and 4 are inflection points.

about the line of symmetry, which in this case is $y = -\frac{1}{2}$. Figure 4.5 shows the tangent lines at each of the four finite points. The tangent line at any such point P intersects the curve in the negative of $2P$, so $2P$ itself is the reflection about $y = -\frac{1}{2}$. If we start from 1 and successively add 1, our routine leads to the figure's labeled group elements $1 \to 2 \to 3 \to 4$. The y-axis through 1 and 4 brings us to 0, completing the group $\mathbb{Z}_5$. ◇

Example 4.1.7. It's well known that for integers $n \geq 3$, $a^n + b^n = c^n$ cannot be solved using only nonzero integers a, b, c — that's Fermat's Last Theorem. For $n = 3$, the associated algebraic curve is $x^3 + y^3 = 1$ and correspondingly, the only rational points on this elliptic curve (besides the point at infinity) are the trivial points $(1, 0)$ and $(0, 1)$. However, in the world of elliptic curves, these trivial points take center stage, meaning that one may never think of Fermat's Last Theorem for $n = 3$ the same way again! Figure 4.6 shows the affine and disk views of $x^3 + y^3 = 1$. In the disk view, the three rational points are labeled 0, 1, and 2. One can check that all three points are inflection points. Therefore the line through any two

88   Chapter 4. Every Elliptic Curve Is a Group!

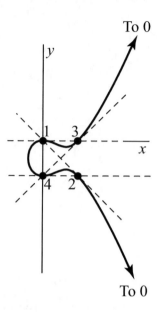

Figure 4.5. The rational points of this elliptic curve form the cyclic group $\mathbb{Z}_5$, with the labeling showing one possible mapping of $\mathbb{Z}_5$ onto the curve's five rational points.

of the three points at, say, $(1,0)$ in the affine view intersects the curve once again in $(1,0)$. The line of symmetry is the 45° line through the origin, so the reflection of $(1,0)$ about this line is $(0,1)$. In the disk view, this says $1+1=2$. One can similarly see that $2+2=1$ and that the three points form the cyclic group $\mathbb{Z}_3$. ◇

Example 4.1.8. Figure 4.7 depicts two real outliers. The curve on the left has equation $y^2 = x^3 - 3x$ and happens to have only one rational point in this affine view — the origin. (The other two x-intercepts are irrational: $\pm\sqrt[3]{6}$.) The y-axis is tangent to the curve and intersects it at infinity. If the origin is labeled 1, then $1 + 1 = 0$ (the point at infinity). Since in projective space the curve has only two rational points, these points form the 2-element group $\mathbb{Z}_2$.

Figure 3.1 on p. 56 depicts the curve $3x^3 + 4y^3 + 5 = 0$ which, as it curves around in the real plane, avoids every rational point there. The

4.1. Finite Groups Within an Elliptic Curve

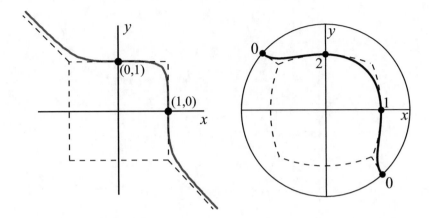

Figure 4.6. This Fermat curve is elliptic and has three rational points. These form the cyclic group $\mathbb{Z}_3$, and the picture's labeling shows one possible one-to-one mapping of $\mathbb{Z}_3$ onto this group. The other possible one-to-one mapping reverses group labels 1 and 2 in the disk view.

curve $y^2 = x^3 + 6$ on the right in Figure 4.7 does the same and furthermore is in Weierstrass short form, so the point at infinity is a triple point and serves as the 0-element in the group consisting of only that zero element. ◊

Exercise 4.1.9. In Figure 4.1 on p. 83, the labels $(0,1)$ and $(1,0)$ can be switched to give the four points a group structure in a different way. Is this the only possibility?

Exercise 4.1.10. Verify that the six points in the projective view in Figure 4.4 in fact form the group $\mathbb{Z}_6$.

Exercise 4.1.11. Draw the appropriate projective line(s) in Figure 4.6 showing that $0 + n = n$ for $n = 0, 1, 2$.

Exercise 4.1.12. Are the two "untilted" curves in Exercise 1.9.6 on p. 29 elliptic? Does each have three rational points? Is each a point of inflection?

Comment 4.1.13. In Example 4.1.6, we met an important curve given not in Weierstrass short form, but in the long form

$$y^2 + a_1 xy + a_3 y = x^3 + a_2 x^2 + a_4 x + a_6. \tag{4.1}$$

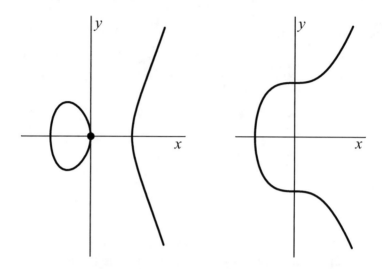

Figure 4.7. The curve $y^2 = x^3 - 3x$ on the left has two rational points in the projective disk and has group $\mathbb{Z}_2$, while the curve $y^2 = x^3 + 6$ on the right has only the point at infinity as rational point, so its group is the trivial 1-element group $\{0\}$, denoted $\mathbb{Z}_1$.

In fact, mathematicians have created an extremely large database of elliptic curves; it is freely accessible at www.LMFDB.org, and extries in it are given in Weierstrass long form, not short. Long-form coefficients are often far smaller than short-form coefficients and are never larger in size than short form coefficients. Long-form constants are entered in LMFDB in the order $[a_1, a_2, a_3, a_4, a_6]$.   ◇

4.2 Finitely Generated Abelian Groups

In the previous section we met one of the central ideas of elliptic curves — that their rational points form an abelian group. We looked at only curves having finitely many rational points and groups whose elements consist of those rational points. That meant, of course, that each group is finite. We are about to meet elliptic curves having infinitely many rational points. At this point Mordell's Theorem becomes a real kingmaker. His result bears repeating:

4.2. Finitely Generated Abelian Groups

> On any elliptic curve there exists a finite set of rational points so that starting from this set, the connect-the-dots routine can produce any rational point on the curve.

The key word here is "finite." It means that the connect-the-dots routine requires only a finite number of starting points to generate all of the elliptic curve's rational points, and this in turn means that the associated group is *finitely generated*. We know even more: Geometrically, two points determine the same line in no matter what order the points are taken. Therefore:

> The group of all rational points on any elliptic curve is a *finitely generated abelian group*.

There's a basic theorem saying what any such group looks like, and that's what this short section is about.

Notation 4.2.1.

- As usual, $\mathbb{Z}$ denotes the set of all integers, positive, negative, and zero. $\mathbb{Z}^r$ denotes the direct sum of r copies of $\mathbb{Z}$, which can also be thought of as the set of all r-tuples $(a_1, a_2, \ldots, a_r)$, $a_i \in \mathbb{Z}$. In a direct sum, addition is always performed componentwise in this book.

- $\mathbb{Z}_m$ denotes the group of integers (mod m) — essentially the integers $0, 1, 2, \ldots, m-1$ supplied with "clock arithmetic" — that is to say, addition (mod m). This group is *cyclic*. $\mathbb{Z}_m^s$ denotes the direct sum of s copies of $\mathbb{Z}_m$. This can also be thought of as the set of all s-tuples $(a_1, a_2, \ldots, a_s)$, $a_i \in \mathbb{Z}_m$.

- If m in $\mathbb{Z}_m$ is a prime power p^s (s positive), then $\mathbb{Z}_m$ is called a *prime power cyclic group*, or simply a *prime power cyclic*, and we write $\mathbb{Z}_{p^s}$. In self-explanatory notation, the *direct sum of prime power cyclics* is denoted $\mathbb{Z}_{p_1^{s_1}} \oplus \cdots \oplus \mathbb{Z}_{p_n^{s_n}}$. The p_i need not be distinct.

- A group is *finitely generated* if there are finitely many elements $\{e_i\}$ in the group so that any group element is a finite sum of various (and possibly repeated) e_i. ◇

Here's the big result:

Theorem 4.2.2.

> Any finitely generated abelian group is a direct sum
> $$\mathbb{Z}^r \oplus \mathbb{Z}_{p_1^{s_1}} \oplus \cdots \oplus \mathbb{Z}_{p_n^{s_n}},$$
> where the $\mathbb{Z}_{p_i^{s_i}}$ may be repeated and where r and/or the number of occurences of $\mathbb{Z}_{p_i^{s_i}}$ may be zero.

For a proof, see [Hungerford, Chapter 2]. If $r = 0$ in $\mathbb{Z}^r$, then there are no copies of $\mathbb{Z}$ and the group is finite.

Definition 4.2.3. The integer r in Theorem 4.2.2 above is called the *rank* of the finitely generated abelian group. The summand $\mathbb{Z}_{p_1^{s_1}} \oplus \cdots \oplus \mathbb{Z}_{p_n^{s_n}}$ in the theorem is called the *torsion part* of the group. ◇

Exercise 4.2.4. Let
$$G = \mathbb{Z}_{p_1} \oplus \mathbb{Z}_{p_2} \oplus \cdots \oplus \mathbb{Z}_{p_n},$$
where all primes p_i are distinct. Prove that G is isomorphic to the cyclic group of order $p_1 \times \cdots \times p_n$.

Exercise 4.2.5. There are exactly six nonisomorphic abelian groups having 108 elements. Find them.

Exercise 4.2.6. True or false? If the group of an elliptic curve
$$y^2 = x^3 + ax + b$$
is $\mathbb{Z}_p$ where p is prime, then its discriminant must be negative. Justify your answer.

4.3 Rank

From what we said in the boxes on p. 91, the set of all rational points of any elliptic curve in $\mathbb{P}^2(\mathbb{R})$ forms a finitely generated abelian group, so Theorem 4.2.2 tells us that the group has the form
$$\mathbb{Z}^r \oplus \mathbb{Z}_{p_1^{s_1}} \oplus \cdots \oplus \mathbb{Z}_{p_n^{s_n}}.$$

Definition 4.3.1. The above nonnegative integer r is called the *rank* of the elliptic curve. ◇

4.3. Rank

Comment 4.3.2. We will see in the next section that for the group of rational points of an elliptic curve, the summand $\mathbb{Z}_{p_1^{s_1}} \oplus \cdots \oplus \mathbb{Z}_{p_n^{s_n}}$ is actually quite special — there are only 15 different possibilities. Stay tuned! ◊

As simple as the notion of rank is, there's a basic fact of life about it that accounts for much of today's effort toward understanding elliptic curves. In fact, questions about rank are today one of the most active and elusive areas in the world of elliptic curves, so let's highlight this reality:

> For the vast majority of questions about the rank of elliptic curves, the answer is unknown. There are lots of reasonable conjectures and lots of data, but few proofs.

Examples:
- Given a particular elliptic curve, what is its rank? Is there some algorithm — some finite sequence of steps — to find out? Around 1960, two British mathematicians, Bryan Birch and Peter Swinnerton-Dyer, formulated a conjecture that has proved to be central and one that most workers in the field think is true. As of this writing, there is still no full answer, but it is so important and so many purported results hinge on it that finding a proof has become one the Clay Institute's million-dollar challenges. We explore this important conjecture in Chapter 5.
- If we take a huge basket of elliptic curves and choose a curve at random, what is the probability that it will be of rank 0? Data based on some 100 million elliptic curves suggest to some that the average rank approaches at least 0.87. But other experts think the answer is 0.5, as well as 0.5 for the probability that the rank is 1. This means rank > 1 is rare enough that the average rank of all elliptic curves is 0.5. Today, this story is still evolving.
- Many researchers believe that there are curves of arbitrarily high rank. Experience so far has led mathematicians to believe that the higher the rank, the rarer the curve — a very reasonable conjecture but not proved as of this writing. However, a 2019 paper ([PPVW]) offers heuristics and calculations suggesting that there are but finitely many elliptic curves of rank greater than 21.
- There's an online database for elliptic curves at www.LMFDB.org listing the rank and other vital statistics for a large number of them, suggesting there's a theorem out there. (For the curious, the abbreviation

LMFDB stands for "L-functions and Modular Forms DataBase." See [LMFDB].) But exactly what that theorem is and how it might be proved are beyond our present reach. Data collected so far indicates that ranks beyond 1 become increasingly scarce. Even finding curves of moderately high rank can prove extremely challenging. As of this book's writing, the highest rank ever found is at least 28, and it's due to Noam Elkies of Harvard. The coefficients of this curve are monstrously large:

$$y^2 + xy + y = x^3 - x^2 - 20{,}067{,}762{,}415{,}575{,}526{,}585{,}033{,}208{,}209{,}338{,}\\542{,}750{,}930{,}230{,}312{,}178{,}956{,}502 x\\ + 34{,}481{,}611{,}795{,}030{,}556{,}467{,}032{,}985{,}690{,}390{,}720{,}374{,}\\855{,}944{,}359{,}319{,}180{,}361{,}266{,}008{,}296{,}291{,}\\939{,}448{,}732{,}243{,}429.$$

Notice that this equation is in Weierstrass long form. It can be put in short form, but then the coefficients would be even larger!

- Given how increasing the rank seems to send us into ever rarer atmosphere, it is reasonable to ask about pushing to an extreme. Are there elliptic curves of *infinite* rank? Intuition would suggest that the atmosphere thins out to nothing — that there are no such curves. This intuition is exactly right — it's a basic implication of the finiteness part of Mordell's Theorem on p. 80!

4.4 Mazur's Theorem

On p. 93 we offered a good dose of reality:

> For most questions about rank, the answer is unknown.

Mordell's Theorem, however, is an outstanding exception which led in a natural way to the structure theorem telling us that the set of rational points of any elliptic curve is a direct sum $\mathbb{Z}^r \oplus \mathbb{Z}_{p_1^{s_1}} \oplus \cdots \oplus \mathbb{Z}_{p_n^{s_n}}$. We suggested in Comment 4.3.2 on p. 93 that the actual torsion part of an elliptic curve's group is not at all as general as $\mathbb{Z}_{p_1^{s_1}} \oplus \cdots \oplus \mathbb{Z}_{p_n^{s_n}}$. The answer comes in the form of another breakthrough, Mazur's Torsion Theorem. Originally conjectured by Oystein Ore and proved in 1977 by Barry Mazur, it tells the story about the torsion part of any elliptic curve's group of rational points.

4.4. Mazur's Theorem

Theorem 4.4.1 (Mazur's Torsion Theorem).

> Of the infinitely many different finite abelian groups, only 15 of them can ever be the torsion part of the group of rational points of an elliptic curve. Eleven of these 15 are pure cyclic, and the remaining four are $\mathbb{Z}_2$ times the simplest even cyclics.
>
> The 15 possible torsion parts are:
>
> - The eleven cyclics $\mathbb{Z}_n$ from $n = 1$ to 12, excluding $\mathbb{Z}_{11}$.
> - The other four are $\mathbb{Z}_2 \oplus \mathbb{Z}_2$, $\mathbb{Z}_2 \oplus \mathbb{Z}_4$, $\mathbb{Z}_2 \oplus \mathbb{Z}_6$, and $\mathbb{Z}_2 \oplus \mathbb{Z}_8$.

The proof of his theorem is long and intricate. For those with sufficient background, a proof appears in [Mazur].

Comment 4.4.2. $\mathbb{Z}_1$ is the cyclic group of one element — the 0-group. In the context of Mazur's Theorem, it is called "the trivial torsion group." ◊

Mazur's Theorem implies that we could look at a hundred billion elliptic curves — or a googleplex to the googleplex power of them — but we'd never once encounter a curve with $\mathbb{Z}_{11}$ as a subgroup, or $\mathbb{Z}_{13}$, $\mathbb{Z}_{14}$, $\mathbb{Z}_{15}$, ... Nor would we ever find, say, $\mathbb{Z}_4 \oplus \mathbb{Z}_4$ as a subgroup. Those 15 torsion groups represent a full accounting of all possible torsion subgroups of rational points in elliptic curves. There is a certain kinship here with our observation that curves with higher rank get harder to find, disappearing completely when we ask for infinite rank. Torsion groups, too, tend to get rarer as we ask for more points in them. The very rarely encountered 16-element group $\mathbb{Z}_2 \oplus \mathbb{Z}_8$ is the largest possible torsion group on any elliptic curve.

Comment 4.4.3. You can check that the large online database for elliptic curves LMFDB.org mentioned on p. 93 lists 1,683,021 curves having only the 0-group as torsion group. This group corresponds to $n = 1$ in the box stating Mazur's Theorem. A sketch of just one such elliptic curve appears on the right in Figure 4.7 on p. 90. There are many other curves having simple Weierstrass forms. Three examples are $y^2 = x^3 - 16$, $y^2 = x^3 - 4x - 4$, and $y^2 + y = x^3 - x - 1$. ◇

In the next section, we add a visual element to Mazur's Theorem. The group of rational points of any elliptic curve is the direct sum of $\mathbb{Z}^r$ ($r \geq 0$) and a (trivial or nontrivial) torsion part. Choose a generator for each summand (a finite cyclic or $\mathbb{Z}$). If the summand is finite, the generator point is denoted by a round dot. If the summand is $\mathbb{Z}$, the point is represented by a square dot. In terms of these dots, it is commonly believed that there's no upper bound to the number of square dots in a plot on the curve, although there will never be infinitely many of them. But the story of round dots is completely different. Mazur's Theorem tells us that there must be either 0, 1, or 2 round dots. If there are no round dots, there's no nontrivial torsion part to the curve's group, making the group the direct sum of r copies of $\mathbb{Z}$ for some $r \geq 0$. If there's just one round dot, then the curve's finite group is cyclic, with four ways it can be of nontrivial odd order (3, 5, 7, or 9) and six ways it can be of even order (2, 4, 6, 8, 10, or 12). The theorem also says that there can never be more than two round dots, and any such pair arises courtesy of the above four direct products with $\mathbb{Z}_2$. Peeking ahead to the table on p. 101, we see that the table suggests that if there is a pair, it's a good bet that the finite group is the smallest of its kind, $\mathbb{Z}_2 \oplus \mathbb{Z}_2$.

In the following exercises, it will be helpful to consult LMFDB.org.

Exercise 4.4.4. The group of the elliptic curve

$$y^2 = x^3 - 91x - 90$$

is $\mathbb{Z}_2 \oplus \mathbb{Z}_2$, so the group requires two generators. Find two different labelings of the four points reflecting its group structure.

Exercise 4.4.5. The group G of the elliptic curve

$$y^2 + xy + y = x^3 - 19x + 26$$

has generators $(3, -2)$ and $(-2, 8)$. What is the structure of G? How many other elliptic curves in the LMFDB database have the same group structure?

4.5. A Gallery of Positive Rank Elliptic Curves

Exercise 4.4.6. The group of the elliptic curve

$$y^2 + xy = x^3 - 4x - 1$$

has eight elements. Three of them are the identity, $(-2, 1)$ and $(5, 8)$. Find the other five.

4.5 A Gallery of Positive Rank Elliptic Curves

In all the examples starting with the one on p. 82, the elliptic curve has rank 0. It's now time to look at some curves of positive rank. Figures 4.8 on p. 98 and 4.9 on p. 99 offer eight examples culled from LMFDB.org. Also, see Exercise 4.7.3 on p. 103.

A positive rank means its group can be $\mathbb{Z}^r$ for some positive r, but we also know that the group can look like, say, $\mathbb{Z}^r \oplus \mathbb{Z}_2$ or $\mathbb{Z}^r \oplus \mathbb{Z}_2 \oplus \mathbb{Z}_4$, where the group has some nontrivial torsion subgroup. In Figures 4.8 and 4.9 we meet curves of rank 1, 2, and 3. The dots on each curve give the locations of group generators.

We've seen that if P is known to be a generating point on an elliptic curve defined by $y^2 = x^3 + ax + b$, it generates a group via the curve's tangent line at P, from which $P + P$ can be constructed. Through repeatedly adding P, we construct nP, and by reflecting nP about the x-axis we get $-nP$. If this procedure unendingly generates new points on the curve, then it creates a copy of the group $\mathbb{Z}$. If at some stage it ceases making new points, then the procedure has created a finite group.

Let's see how this idea works for the curve $y^2 = x^3 - x + 1$, which is (a) in Figure 4.8. The generator is $P = (1, -1)$, and one can check that its tangent line passes through $(-1, 1)$ in the curve and has reflection $2P = (-1, -1)$. For the next iteration, the line through P and $2P$ passes through $(0, -1)$ and reflects to $P + 2P = 3P = (0, 1)$. Here are a few successive iterations: $4P = (3, 5)$, $5P = (5, -11)$, $6P = (\frac{1}{4}, -\frac{7}{8})$, $7P = (-\frac{11}{9}, -\frac{17}{27})$.

The curve $y^2 = x^3 + 2x + 3$ in (b) of Figure 4.8 has two generators, one being $(3, 6)$ which generates $\mathbb{Z}$, and $(-1, 0)$ which generates $\mathbb{Z}_2$. This last is easy to see, since if P is $(-1, 0)$, then $2P$ is where the tangent to the curve at $(-1, 0)$ intersects the curve, and that is the point at infinity at the ends of the y-axis, so $2P$ is the 0-element of the group $\{0, P\} = \mathbb{Z}_2$ generated by P. From the standpoint of complexity, the group generated by $(3, 6)$ is a different animal. The tangent line to the curve at P intersects the curve

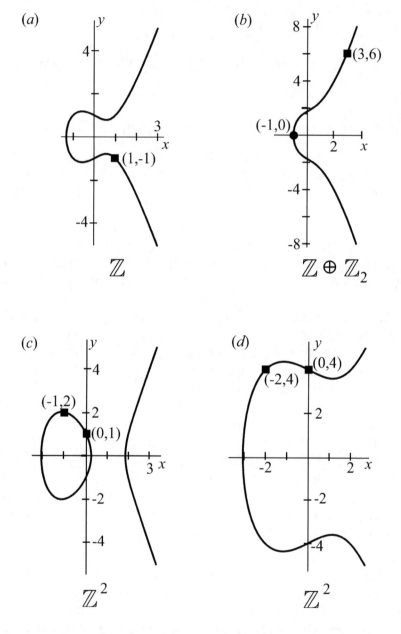

Figure 4.8. (a): $y^2 = x^3 - x + 1$ is a rank 1 curve with no torsion, having generator $(1, -1)$. (b): $y^2 = x^3 + 2x + 3$ has rank 1 with torsion — $(3, 6)$ generates $\mathbb{Z}$ and $(-1, 0)$ generates $\mathbb{Z}_2$. (c) and (d): $y^2 = x^3 - 4x + 1$ and $y^2 = x^3 - 4x + 16$ are both torsion-free of rank 2; each curve has group $\mathbb{Z}^2$.

4.5. A Gallery of Positive Rank Elliptic Curves

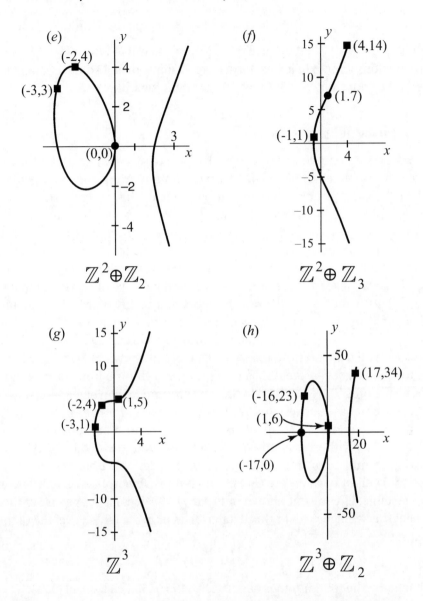

Figure 4.9. (e): $y^2 + xy = x^3 + x^2 - 6x$ is mixed of rank 2. $(0,0)$ and $\{\infty\}$ make up the torsion part, so this curve's group is $\mathbb{Z}^2 \oplus \mathbb{Z}_2$. (f): The curve $y^2 = x^3 + x^2 + 23x + 24$ has rank 2, with torsion part $\mathbb{Z}_3$ generated by $(1, 7)$. (g): This curve $y^2 = x^3 + x^2 + x + 22$ has pure rank 3. (h): The curve $y^2 = x^3 + x^2 - 255x + 289$ has rank 3. The 2-element torsion group is generated by $(-17, 0)$, with the other element on the line at infinity.

at $(-\frac{23}{144}, \frac{-2{,}827}{1{,}728})$, so $2P$ is $(-\frac{23}{144}, \frac{-2{,}827}{1{,}728})$. The fractions quickly grow in complexity, with $3P = -(\frac{193{,}101}{207{,}025}, -\frac{53{,}536{,}482}{94{,}196{,}375})$. At just $7P$, the coordinates are fractions which, in lowest terms, are 92 digits over 93 digits. (Consider what the coordinates of $100P$ or $10^{10}P$ might look like!)

4.6 How Many Curves?

It was easy to find curves of low rank and having small or trivial torsion like those depicted in Figure 4.8. Curves of higher rank and/or larger torsion groups are rarer and their equations typically have larger coefficients. Notice that not a single equation in Figure 4.9 is in the short Weierstrass form $y^2 = x^3 + ax + b$, and that's because forcing an equation into this form comes at the cost of larger coefficients. Intuitively, if we randomly select an elliptic curve from a large database of them, the chances are quite good that we'll get a curve of rank 0 or 1, and perhaps only trivial torsion. Even slightly increasing the rank and torsion quickly puts us in more rarefied territory. How rare, for example, is a curve with group $\mathbb{Z}^3 \oplus \mathbb{Z}_4$ — a curve of rank 3 with cyclic torsion of order 4? We can get a sense of how ranks are distributed from the large database at www.LMFDB.org which, as this book goes to press, has more than 3 million elliptic curves in it. Using the procedure outlined in the next section, we can determine that of all these database entries, there are only four curves having group $\mathbb{Z}^3 \oplus \mathbb{Z}_4$. What about assuming a rank of just 1 but asking for more torsion, such as $\mathbb{Z}_2 \oplus \mathbb{Z}_8$? Now there are only three such curves in the database. A small change in rank and/or torsion can strongly affect how unusual the corresponding elliptic curves are. But compared to a few isolated examples, an actual table, like the one shown in Figure 4.10, gives a better sense of this. To get the average rank in this database, take the weighted average of the last row:

$$0 \cdot 956{,}089 + 1 \cdot 1{,}246{,}540 + 2 \cdot 274{,}346 + 3 \cdot 6{,}679 + 4 \cdot 1$$

divided by the total number of entries, $2{,}483{,}655$. This comes to

$$\frac{1{,}815{,}273}{2{,}483{,}655} \approx 0.73.$$

This average doesn't lend a lot of credibility to the belief, held for many years, that the limiting average is one-half. However, this isn't the end of the story! We'll continue the tale in Section 5.5.

4.7. Finding Generators

Rank	0	1	2	3	4
Torsion-free	477,703	679,194	186,326	6,018	1
$\mathbb{Z}_2$	404,596	485,681	78,532	635	0
$\mathbb{Z}_3$	19,141	21,559	2,875	11	0
$\mathbb{Z}_4$	13,118	13,985	1,538	4	0
$\mathbb{Z}_5$	612	638	59	0	0
$\mathbb{Z}_6$	2,807	2,913	225	0	0
$\mathbb{Z}_7$	40	36	0	0	0
$\mathbb{Z}_8$	94	66	2	0	0
$\mathbb{Z}_9$	10	8	0	0	0
$\mathbb{Z}_{10}$	26	16	0	0	0
$\mathbb{Z}_{12}$	8	8	0	0	0
$\mathbb{Z}_2 \oplus \mathbb{Z}_2$	37,147	41,655	4,729	11	0
$\mathbb{Z}_2 \oplus \mathbb{Z}_4$	730	734	59	0	0
$\mathbb{Z}_2 \oplus \mathbb{Z}_6$	51	41	1	0	0
$\mathbb{Z}_2 \oplus \mathbb{Z}_8$	3	3	0	0	0
Σ	956,089	1,246,540	274,346	6,679	1

Figure 4.10. Rank and torsion data for about $2\frac{1}{2}$ million elliptic curves.

Exercise 4.6.1. What percentage of curves in Figure 4.10 requires exactly three generators for the group?

4.7 Finding Generators

In Section 1.12 starting on p. 33, we addressed a natural question: If we know, say from Legendre's Criterion, that there's a rational point on a smooth conic (and we need such a point to make our rational-slope lines method work), how do we find it? In Appendix C we provide search code that works nicely when the quadratic's coefficients lie between about -30 and $+30$, and [Cremona] provides an approach for much larger coefficients. This, as well as an algorithm for finding a rational point on an ellipse, parabola, or hyperbola defined over $\mathbb{Q}$ were briefly alluded to on pp. 33–34.

We now face a similar situation with elliptic curves — we need rational starting points for our connect-the-dots method. But how do we find them? It turns out that even the Hasse-Minkowski Theorem doesn't help us here. We *could* adapt our search code by replacing x and y in $y^2 = x^3 + Ax + B$ by $\frac{a}{c}$ and $\frac{b}{c}$ and multiplying through by c^3 to get a homogeneous cubic in a, b, and c. Any nonempty list of triples (a, b, c) the code produces leads to rational points $x = \frac{a}{c}$ and $y = \frac{b}{c}$ guaranteed to satisfy $y^2 = x^3 + Ax + B$. The trouble is that such a list of rational points doesn't do us much good. We don't need just *some* rational point, we need a generating one. Furthermore, the curve may need several generating points, and our list mixes all the rational points together — if the rank is 2 and we choose two points, both of them might belong to the sequence arising from the same generator. In short, our code applied to elliptic curves isn't of much help.

Fortunately, mathematicians have used theory, the Birch–Swinnerton-Dyer Conjecture which we'll meet in the next chapter, and a lot of computer work to come up with generators for many elliptic curves, and the fruits of this big undertaking are available online. As of this writing, the database includes information for 3,064,705 curves. Here are steps leading to this information, including concrete generators for any elliptic curve in the database.

- **Enter www.LMFDB.org/ into a search engine such as Google.**

- **On the computer screen there will appear headings such as "Introduction and more", "L-functions", "Modular Forms", "Varieties." Go to "Varieties" and select "Elliptic /$\mathbb{Q}$".**

- **This will display the page where the curve's constants are entered and will accept a, b in the short form y² = x³ + ax + b or a_1, a_2, a_3, a_4, a_6 in the long form**

$$y^2 + a_1xy + a_3y = x^3 + a_2x^2 + a_4x + a_6.$$

The entry box will probably contain something like 11.a2, but just go ahead and type $-1, 1$ or, if you want, $[-1, 1]$; either one will specify the curve $y^2 = x^3 - x + 1$. Similarly enter integers a_1, a_2, a_3, a_4, a_6 with or without brackets for an equation in long form. The page will respond by showing the curve's equation, the group structure, the generators of infinite order, and any torsion-group generators.

4.7. Finding Generators

The page includes many other results, but for the purposes of this book, these are basic ones we're interested in.

- The program also displays a plot of the elliptic curve, its discriminant, rank, and torsion structure. Right click on the plot and in the drop-down menu choose "Open link in new window"; this leads to a much larger and nicer picture of the curve. When data for various curves are entered, notice how the discriminant is indeed negative for curves having only a branch and positive for those with both branch and oval.

Example 4.7.1. As an example, for the elliptic curve $y^2 = x^3 + 2x + 3$, enter 2, 3 in the entry box. The program immediately displays the curve's group: $\mathbb{Z} \oplus \mathbb{Z}_2$, so the curve's group has rank 1 and a torsion subgroup of order 2. Just below that, it tells us that $P = (3, 6)$ is a generator of $\mathbb{Z}$ in $\mathbb{Z} \oplus \mathbb{Z}_2$ and $(-1, 0)$ is a generator of the torsion subgroup $\mathbb{Z}_2$. ◊

Example 4.7.2. As an example of entering long-form data, type in

$$0, 1, 0, 23, 24.$$

The response screen shows the curve's equation

$$y^2 = x^3 + x^2 + 23x + 24$$

as well as the curve's group $\mathbb{Z}^2 \oplus \mathbb{Z}_3$. Below that we see two infinite-order generators $(-1, 1)$ and $(4, 14)$ and just below that, a generator $(1, 7)$ of the torsion subgroup $\mathbb{Z}_3$. ◊

Exercise 4.7.3. Use LMFDB to verify the generators given in the curves depicted in Figures 4.8 and 4.9.

Exercise 4.7.4. Find the long-form equation for the curve $y^2 = x^3 + ax + b$ after it's been translated upward by one unit.

5

A Million-Dollar Challenge

5.1 Breaking Up a Task into Many Smaller Ones

One of the gaping holes in number theory is that we know of no way to decide in finitely many steps just what the rank of an elliptic curve is. There *is* a conjecture about this — the Birch and Swinnerton-Dyer Conjecture — and if it turns out to be true, then we would have such a way. Around 1960, Bryan Birch and Peter Swinnerton-Dyer did some computations of ranks of elliptic curves, not by counting rational points on the curve itself (that's actually very difficult) but rather by breaking up that big task into lots of smaller pieces. Each small task consists in counting the rational points on a greatly simplified version of the original curve, each version corresponding to a different prime p. In official language, each simpler curve is a "reduction of the original curve (mod p)." (We'll explain what this means in a moment.) Each such reduced curve will have only finitely many points, and a computer can automatically find what that finite number is. By working through successive runs of primes and counting how many points there are in each corresponding reduced curve, the computer can build up a statistical profile for the original curve. To get a good profile, the computer must do this up to really large primes — primes much larger than a million. For such large runs of primes, Birch and Swinnerton-Dyer conjectured that as the runs get longer and longer, the evolving profile of an elliptic curve defined over $\mathbb{Q}$ looks more and more like a line whose slope is an integer. Specifically, if that line has slope 1, the curve's rank is 1, and if a curve's plotted profile converges to a line of slope 2, then the curve's rank is 2, and so on.

To get the gist of these ideas in a more concrete way, we now illustrate reducing a specific curve modulo a specific prime. For this, we choose a curve, then a prime p, then show how to carry out reducing that curve (mod p). We finally actually plot our reduced curve.

- We choose the curve $y^2 = x^3 - 25x$. This is similar to $y^2 = x^3 - x$ which we've depicted in Figures 3.16 and 4.1, but $y^2 = x^3 - 25x$ has zeros at 0, ±5 instead of 0, ±1. This choice will keep things nontrivial yet simple.
- Second, continuing to keep things nontrivial and easy to understand, we choose the prime p to be 29.
- Each of the curve's coefficients can be taken to be integers, and we reduce the curve (mod 29) by reducing each of those coefficients (mod 29). In our case of $y^2 = x^3 - 25x$, -25 (mod 29) is $-25 + 29 = 4$, so the equation reduces to $y^2 = x^3 + 4x$ (mod 29).
- Finally, we plot $y^2 = x^3 + 4x$ (mod 29). This plot isn't on standard "graph paper" $\mathbb{R}^2$ or even $\mathbb{Q}^2$. Rather, the plotting is done in a far smaller world — in our case, $\mathbb{Z}_{29}^2$. This "graph paper" has only a finite number of points on it (29^2 points). We begin by plotting just the right-hand side, $x^3 + 4x$ (mod 29). Now $x^3 + 4x$ is a function, so a value is produced for each $x \in \mathbb{Z}_{29}$. In Figure 5.1, we've represented each point (x, y) by a small hollow circle. The plot of $y^2 = x^3 + 4x$ (mod 29) appears in $\mathbb{Z}_{29}^2$ as a collection of solid small disks or dots. Notice that sometimes there are (and other times there are not) solid dots in the column containing a hollow circle. There's a reason for this: Some elements of $\mathbb{Z}_{29}$ possess square roots, and others don't. For example, 5 does and 10 doesn't. ($5 = 11^2 = 18^2$, but square every element of $\mathbb{Z}_{29}$ and you'll never get 10.) To plot y-values, $x^3 + 4x$ (which is y^2) must have square roots! So there are solid dots exactly when the plotted value of $x^3 + 4x$ is a square (mod 29). For convenience, in Figure 5.1, on the y-axis, each square value in $\mathbb{Z}_{29}$ has a little square drawn around it.

Comment 5.1.1. It's easy to find $x^3 + 4x$ (mod 29) when x is small — say, 1, 2, 3, 4. But evaluating $x^3 + 4x$ (mod 29) for larger values of x requires more computation, and that can become tedious when done by hand. There's an easier way. In a browser, enter

$$\mathtt{WolframAlpha : ComputationalIntelligence}$$

5.1. Breaking Up a Task into Many Smaller Ones

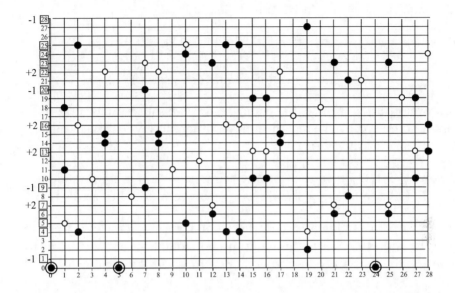

Figure 5.1. The set of 39 solid dots makes up the reduction modulo 29 of the elliptic curve $y^2 = (x + 5)x(x - 5)$. In the next section we explain further aspects of this figure.

This brings up a free symbolic manipulation package which, among other things, makes it easy to evaluate any integer modulo another integer. For example, to evaluate $x^3 + 4x \pmod{29}$ when x is, say, 23, enter

$$\texttt{solve } (23)^3 + 4 * (23) \bmod 29.$$

WolframAlpha at once returns the answer, 21. For $x = 8$, say, replace each (23) by (8) to get 22. In this way, it becomes easy to plot the values of $x^3 + 4x \pmod p$ in a figure such as Figure 5.1. It's equally easy to plot the solutions to $y^2 = x^3 + 4x \pmod p$. For example, enter

$$\texttt{solve y}^2 = (12)^3 + 4 * (12) \bmod 29.$$

This produces 6 and 23 — the two values of y lying above $x = 12$ satisfying $y^2 = x^3 + 4x \pmod{29}$ and plotted there as solid dots. But in Figure 5.1, there are no solid dots above $x = 11$. The value of $x^3 + 4x \pmod{29}$ above that x is 12, and 12 has no square roots in $\mathbb{Z}_{29}$ — that is, there's no y in

$\mathbb{Z}_{29}$ satisfying $y^2 = 12$. In WolframAlpha we'll see

```
solve y² = (11)³ + 4 * (11) mod 29
```

(no integer solutions exist). ◇

Comment 5.1.2. Notice that 6 and 23, the two square roots of 12 in the above comment, can be regarded in two different ways — as $\pm 6 \pmod{29}$ as well as $\pm 23 \pmod{29}$. ◇

Exercise 5.1.3. Prove that for any prime $p \neq 2$, if a nonzero element $n \in \mathbb{Z}_p$ has a square root, then it has two of them, and they are distinct.

Exercise 5.1.4. In Figure 5.1, the horizontal line of symmetry is $y = 14.5$. Find examples of three points lying on a horizontal line and verify that the connect-the-dots procedure works for each choice of three dots.

Exercise 5.1.5. Repeat the above exercise for three points lying on a non-horizontal line.

Comment 5.1.6. If Maple is used to work out Exercise C.1 on p. 199, the code used can be easily modified to work modulo any prime. Ending each line of Maple code with `mod 29;` (including the semicolon) tells Maple to track each computation modulo 29. Starting from the integral point $(45, 300)$ on $y^2 = x^3 - 25x$, apply the modulo code to Figure 5.1 until the program says the calculation is impossible, signaling that the line's slope is infinite. That means the next point is at infinity. Of course in all this section, 29 is but one choice. Arbitrarily pick the prime $p = 153$, say, and the reduction turns out to consist of 172 points, with 171 of these in the "graph paper" $\mathbb{Z}_{153} \oplus \mathbb{Z}_{153}$ and one at infinity. Pick $p = 201$ and we happen to get 201 finite points and one at infinity. ◇

Comment 5.1.7. Since we're reducing curves modulo a prime rather than a nonprime, there is a related way of looking at things. When m is a prime p, $\mathbb{Z}_m$ has an important status: $\mathbb{Z}_p$ is also a *field*, meaning that in addition to the group structure, you can not only multiply, you can also divide by any nonzero element, just as in the fields $\mathbb{Q}$, $\mathbb{R}$, and $\mathbb{C}$. For example, if $p = 7$, any fraction having 1, 2, 3, 4, 5, or 6 in the denominator turns out to equal some one of 1, 2, 3, 4, 5, or 6 in $\mathbb{Z}_7$. This doesn't happen for, say, $\mathbb{Z}_6$. For example, when you try to solve for $\frac{1}{3}$ via solving $1 = 3x \pmod{6}$, the

5.1. Breaking Up a Task into Many Smaller Ones

multiples of 3 simply cycle through 0 and 3, never hitting 1. Ditto for $\frac{1}{2}$ — the multiples of 2 cycle through 0, 2, and 4. Arguments of this kind can be made for any m not equal to a prime. A little elementary number theory shows that for any prime p, multiples of any nonzero element of $\mathbb{Z}_p$ cycle through all $p-1$ nonzero elements of $\mathbb{Z}_p$. This has a nice consequence: For any prime p, WolframAlpha can easily evaluate fractions in $\mathbb{Q}$ modulo p. For example, to find $\frac{9}{17}$ in $\mathbb{Z}_{11}$, enter this in WolframAlpha:

```
solve x = 9/17 mod 11
```

and the program promptly returns

$$x = 7.$$

◇

Notation 5.1.8. When $\mathbb{Z}_p$ is considered as a field, we write it as $\mathbb{F}_p$. ◇

Exercise 5.1.9. Construct a 6-by-6 multiplication table for the nonzero elements of $\mathbb{F}_7$. Also, construct a 6-by-6 division table for the nonzero elements of $\mathbb{F}_7$. Now experiment: Using your tables, verify facts such as in $\mathbb{F}_7$, $\frac{1}{2} \times \frac{3}{2} = \frac{3}{4}$, $\frac{-2}{5} \div \frac{3}{4} = -1$, instances of the distributive law, and so on.

Exercise 5.1.10. The cartesian product $\mathbb{F}_p \times \mathbb{F}_p = \mathbb{F}_p^2$ consists of all ordered pairs of elements from $\mathbb{F}_p$, in a way analogous to, say, $\mathbb{Q} \times \mathbb{Q} = \mathbb{Q}^2$ or $\mathbb{R} \times \mathbb{R} = \mathbb{R}^2$. The product $\mathbb{F}_p^2$ can analogously be called "$\mathbb{F}_p$-graph paper." In Figure 5.1 or 5.2, the corresponding graph paper consists of the intersection points of the ruled lines there. Taking a cue from linear algebra, we define a *line through the origin* in $\mathbb{F}_p^2$ to be the set of all integer multiples of some nonorigin point of the graph paper. How many such lines are there in $\mathbb{F}_p^2$? Draw these lines (perhaps using different colored pencils or pens) in $\mathbb{F}_7^2$.

Exercise 5.1.11. We can call $\mathbb{F}_p^2$ an "affine plane," which at once suggests the possibility of an "$\mathbb{F}_p$-projective plane." Describe how to add a point at infinity to the line $\mathbb{F}_7$, with two identified points at infinity creating a "circle." Now projectivize each line through the origin by adding a point at infinity to create $\mathbb{P}^2(\mathbb{F}_7)$. Verify that the number of points at infinity is the same as the number of points in any other projectivized line.

Comment 5.1.12. Extending the above exercise, one can define a general line in $\mathbb{P}^2(\mathbb{F}_p)$ and then establish that any two different lines in $\mathbb{P}^2(\mathbb{F}_p)$ intersect in a single point. ◇

Comment 5.1.13. The above comments and exercises lead to a remarkable aspect of $\mathbb{F}_p$: This field acts very much like $\mathbb{Q}$ in that, for example, the connect-the-dots routine developed for elliptic curves defined over $\mathbb{Q}$ can be carried over to elliptic curves defined over $\mathbb{F}_p$. The analogy holds geometrically using lines, as well as algebraically, and the logical consequences of the method flow out the same way. This means that for points P_1 and P_2 on an elliptic curve over $\mathbb{F}_p$, connecting the dots on P_1 and P_2 to produce P_3 can be interpreted as commutative addition: $P_3 = P_1 + P_2 = P_2 + P_1$. All requirements for a group (Definition 4.1.1) hold — including associativity. The points comprising the curve thus form a finite (hence finitely generated) group, meaning the basic structure result of Theorem 4.2.2 holds. ◇

Comment 5.1.14. The magic in Comment 5.1.13 continues! One can directly reduce modulo a prime all the points of the original curve in $\mathbb{Q}^2$, and that, of course, entails reducing ordered pairs of fractions in $\mathbb{Q}$. Thus a point like $(\frac{1}{5}, \frac{10}{7})$ reduces (mod 11) to (9, 3) by entering

```
solve x = 1/5 mod 11
```
```
solve y = 10/7 mod 11
```

in WolframAlpha. Now the remarkable thing about reducing points in the original curve in $\mathbb{Q}^2$ to points in the curve modulo p is that this reducing operation is a homomorphism from the finitely generated abelian group of the original curve to the abelian group of the curve modulo p. As just one consequence, this means you can add points either before or after reducing. The results are the same. ◇

5.2 The Birch and Swinnerton-Dyer Conjecture

We mentioned earlier that computer data can be used to build up a statistical profile of an elliptic curve's set of rational points and that central to this is breaking up this difficult task into many easier reduced-curve tasks. Birch and Swinnerton-Dyer created a profile in which the various reduced-curve statistics over ever larger runs of primes can be written so that the logarithms of these statistics approximate a straight line with

5.2. The Birch and Swinnerton-Dyer Conjecture

slope equal to the curve's rank. An essential step in their approach can be summed up this way:

> Plot points (x, S), where S is the product of normalized reduced-curve sizes over those primes $p \leq x$.

We now explain just what this statement means.

First, when we reduce a typical elliptic curve (mod p), how many points would we expect its graph in $\mathbb{Z}_p \oplus \mathbb{Z}_p$ to have? Actually, does it even make sense to ask a question like this? The answer is yes. Let's see why. The "x-axis" and the "y-axis" $\mathbb{Z}_p$ each contains p points — 0 through $p-1$. A rational point (x, y) on the curve $y^2 = x^3 + ax + b$, once the point is reduced, gets plotted only if $x^3 + ax + b$ (mod p) happens to be a square in $\mathbb{Z}_p$, since $x^3 + ax + b$ must be y^2. An integer in the vertical axis $\mathbb{Z}_p$ is a square exactly when it equals j^2 for some $j \in \mathbb{Z}_p$. So how many squares *are* there in $\mathbb{Z}_p$? Here's what we need to know:

Theorem 5.2.1.

> For any odd prime p, the number of nonzero squares in $\mathbb{Z}_p$ is $\frac{p-1}{2}$.

A proof of this appears in [Ireland, p. 51] and uses elementary properties of quadratic residues.

Reductions are important only for large p, and a large p is close in a relative sense to $p-1$. We therefore conclude:

> For large p, we can expect that two y-values are plotted for every square integer in the vertical axis $\mathbb{Z}_p$, so the expected number of points in the reduced elliptic curve in $\mathbb{Z}_p \oplus \mathbb{Z}_p$ is p.

Actually, the number of points in reductions of various curves is often different from the expected values. This difference is the result of two opposing "forces" at work. The first represents a decrease from the expected value, and the second, an increase. We will be able to appreciate each of these tendencies through the example illustrated in Figure 5.1 on p. 107.

As we've noted earlier, the solid dots in the figure represent the reduced plot of $y^2 = x^3 - 25x \pmod{29}$, and it's the number of these dots we will be talking about. The expected number of them is $p = 29$, but the actual number turns out to be different from that. In addition to solid black dots in the figure, there are those hollow ones, the values (mod 29) of the function $x^3 + 4x$ as x runs along the horizontal axis, where $x^3 + 4x$ is the reduction of $x^3 - 25x \pmod{29}$. Therefore above each x in the x-axis there's exactly one hollow dot, and the two square roots of the y-value of that hollow dot give two solid dots in that same column in $\mathbb{Z}_{29}^2$. But we saw in the last section that sometimes those square roots don't exist! As an example, we looked at the vertical column above $x = 3$ where there was, as expected, a hollow dot. It has the y-value of 10. But the column has no corresponding square-root solid dots since 10 isn't a square number modulo 29. As an alternative to squaring each element of $\mathbb{Z}_{29}$ and never finding 10, we can simply enter in WolframAlpha

$$\text{solve } y*y = 10 \text{ mod } 29.$$

It responds with

$$\text{(no integer solutions exist)}.$$

It will respond the same way if any nonsquare is entered instead of 10. On the other hand, 5 *is* a square — it's 11^2 as well as 18^2. Enter

$$\text{solve } y*y = 5 \text{ mod } 29$$

and WolframAlpha dutifully responds with

$$y = 11 \pmod{29}; \quad y = 18 \pmod{29}.$$

We saw a moment ago that for any odd prime p, the number of nonzero squares in $\mathbb{Z}_p$ is $\frac{p-1}{2}$ which, with $p = 29$, is 14. Those 14 square numbers appear in the vertical $\mathbb{Z}_{29}$-axis, each surrounded by a square. Now consider a lattice L whose horizontal elements go from 1 to 28 and whose vertical elements consist of the 14 nonzero square integers. (For the vertical side of the lattice, remember that only square numbers have square roots.) Since L has different x and y dimensions, the set of points in L satisfying $(x, x^3 - 25x)$ sets up a competition between increasing and decreasing the solid-dot count away from the expected value p.

In our example, who wins?

- **A fact working toward making the actual value less than the expected value:** The plot in L of $(x, x^3 - 25x)$ is not onto, which is on the

5.2. The Birch and Swinnerton-Dyer Conjecture

side of a decrease. This is because there is no hollow dot for the four square y-values $1, 9, 20, 28$. Those four missing hollow dots mean eight missing solid dots.

- **A fact working toward making the actual value greater than the expected value:** The plot is not one-to-one, which is on the side of an increase — look at the four square y-values $7, 13, 16, 22$. At each of these y-values there is not one, but three x-values, an excess of $(3-1) \times 4 = 8$ hollow dots, or 16 extra solid dots.

- **Who wins?** In total, among points above the bottommost row, there's an excess of $16 - 8 = 8$ solid dots over the expected number of 28 of them. That means $28 + 8 = 36$ solid dots. That, together with three dots at $y = 0$ (the bottommost row), makes for a grand total of 39 points in the reduced plot. So in this case, **increase** wins.

Exercise 5.2.2. Figure 5.2 depicts the graph in $\mathbb{Z}_{11} \oplus \mathbb{Z}_{11}$ of the elliptic curve $y^2 = x^3 - 25x$ reduced modulo $p = 11$. Its reduced equation is therefore $y^2 = x^3 + 8x$. Which integers along the vertical axis are squares (mod 11)? Using WolframAlpha, add to the figure the (mod 11) plot of the function $x^3 + 8x$ using, say, little circles, as well as solid dots for solutions (mod 11) to $y^2 = x^3 + 8x$. As in the $p = 29$ case, find the net gain or loss in solid dots. From this, decide whether the actual number of solid dots agrees or disagrees with the expected number of them.

Let's step back for a moment and consider again the central idea of Birch and Swinnerton-Dyer which is in a box on p. 111:

> Plot points (x, S), where S is the product of normalized reduced-curve sizes over those primes $p \leq x$.

We've just looked at the elliptic curve $y^2 = x^3 - 25x$ and its reduction (mod 29). We argued that the expected number of finite points in its graph is $p = 29$ but that the plotted graph actually has 39 points.

Definition 5.2.3. The *normalized size* of a reduced curve is the actual number N_p of points in the plot divided by p, the expected number of points. ◊

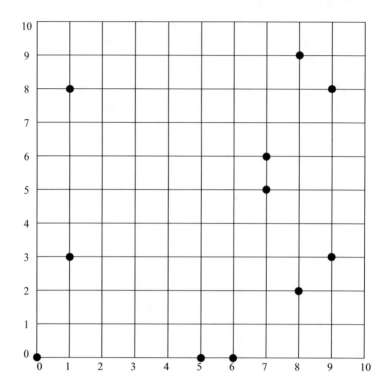

Figure 5.2. This set of solid dots represents the graph in $\mathbb{Z}_{11} \oplus \mathbb{Z}_{11}$ of the reduction of $y^2 = x^3 - 25x \pmod{11}$.

Example 5.2.4. The normalized size of the curve $y^2 = x^3 - 25x \pmod{29}$ is $\frac{N_p}{p} = \frac{39}{29} \approx 1.345$. ◇

Example 5.2.5. The normalized size of the reduced (mod 29) curve defined by $y^2 = x^3 - 25x$ is comparatively large, but at primes 2, 3, 5, 7, and 11, the number of plotted points turns out to be exactly the expected number 2, 3, 5, 7, and 11, respectively, so the normalized size of the curve reduced at each of these five primes is 1. The plot for $p = 13$ consists of 19 points, making its normalized size $\frac{19}{13} \approx 1.46$, while the curve again has 19 points when reduced at $p = 17$, making the normalized size there ≈ 1.12. When reduced at $p = 41$, the curve has relatively fewer points and its normalized size is smaller — about 0.756. Here are approximate normalized sizes for primes up to 59:

5.2. The Birch and Swinnerton-Dyer Conjecture

$p = 2 : 1,$ $\quad p = 3 : 1,$ $\quad p = 5 : 1,$
$p = 7 : 1,$ $\quad p = 11 : 1,$ $\quad p = 13 : 1.46,$
$p = 17 : 1.12,$ $\quad p = 19 : 1,$ $\quad p = 23 : 1,$
$p = 29 : 1.345,$ $\quad p = 31 : 1,$ $\quad p = 37 : 1,$
$p = 41 : 0.756,$ $\quad p = 43 : 1,$ $\quad p = 47 : 1,$
$p = 53 : 1.3,$ $\quad p = 59 : 1.$ $\quad \diamond$

Example 5.2.6. The last box above tells us to plot above each x the product of normalized reduced-curve sizes over those primes $p \leq x$. The data in Example 5.2.5 allows us to do this for $x \in [2, 59)$. For example the primes ≤ 5 all have normalized sizes 1, so their product is 1. In fact, we'd assign the value 1 to every $x < 13$. In the interval $[13,16]$ the assigned value is ≈ 1.46, while in $[17,18]$, it's $\approx 1.46 \times 1.12$, or about 1.64. Farther out, the value given to each x in $[59,60]$ is close to

$$1^5 \times 1.46 \times 1.12 \times 1^2 \times 1.345 \times 1 \times 0.946 \times 0.756 \times 1^2 \times 1.3 \times 1 \approx 2.044 \, . \quad \diamond$$

The plot of x versus the normalized sizes of the curve $y^2 = x^3 - 25x$ for these primes is shown in Figure 5.3. As we see, these values bounce around a bit. This apparent randomness continues for thousands of primes and we see a pattern emerging only when we look at a run of, say, a hundred thousand consecutive primes. At that point, x is nearly 1.3 million. In the early 1960s, Peter Swinnerton-Dyer and Bryan Birch carried out these computations using the EDSAC-2 computer at Cambridge University. (EDSAC is an acronym of Electronic Delay Storage Automatic

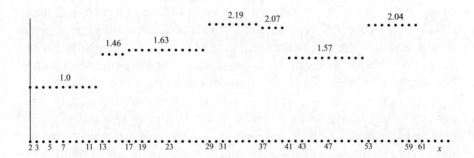

Figure 5.3. This is the plot of normalized sizes of the elliptic curve $y^2 = x^3 - 25x$ for $x = 2, 3, \ldots, 60$.

Calculator.) It was built using arrays of radio tubes that needed constant cooling to prevent failure. The computer had a very limited memory, so programmers had to make every character count. Nevertheless, Swinnerton-Dyer and Birch were able to make a guess. In vastly extending the process in Example 5.2.6 to many thousands of primes and plotting x versus the associated products of thousands of normalized sizes of reduced curves, the evolving shape of the plot seemed to look more and more like some constant K times $\ln(x)$. That guess was associated to rational elliptic curves of rank 1, of which our curve $y^2 = x^3 - 25x$ is an example. When computations and plots were constructed for curves of known rank 2, the products over large runs of primes were on average distinctly bigger than for those of rank 1 curves, and that made sense — rank 2 curves have infinitely many rational points coming from two independent generators, not one, and the graphs seemed to evolve toward a function having larger values. It appeared that some constant K times $(\ln x)^2$ worked. They made a leap of faith and conjectured that for ever larger runs of consecutive primes, the plot of a curve of rank r would asymptotically tend toward $y = K(\ln x)^r$, so for a curve of rank r, their conjecture could be written as

$$\prod_{p \leq x} \frac{N_p}{p} \approx K(\ln x)^r.$$

This is the first important mathematics conjecture arrived at through computer-generated data. Their conjecture was met with skepticism from most researchers, but as time went on and computer power grew, it increasingly appeared that the hunch Birch and Swinnerton-Dyer had was right. Today, virtually all workers in the field believe the conjecture is true. Their guess has been checked numerically on millions of elliptic curves, and *in every one of these millions of curves, the prediction always holds*. As with the Riemann Hypothesis, the Clay Mathematics Institute has offered a million dollars for a proof of this remarkable conjecture.

The right-hand side $K(\ln x)^r$ is a monomial in $\ln x$, and for large values of x its size increases very rapidly with r. With $x = 1,000,000$, say, the base 10 logarithms of $(\ln x)^r$ for $r = 1, 2, 3, 4$ are 6, 36, 216, and 1,296. It's hard to graph these on a single sheet of paper to make meaningful comparisons, but taking the logarithm of the monomial calms things down a lot, turning the plot into merely a line, $Y = mX + b$, where $m = r$, $X = \ln(\ln x)$, and b is a constant. The big standout here is $m = r$, because now the rank of the elliptic curve becomes simply the line's slope.

5.2. The Birch and Swinnerton-Dyer Conjecture

Rational elliptic curves defined by

$$y^2 = x^3 - d^2 x = (x+d)x(x-d)$$

provide examples of curves of rank 0 through 4; see Figure 5.4. This figure shows computer-generated plots of the natural logarithms of

$$\prod_{p \leq x} \frac{N_p}{p}$$

for x increasing to very large numbers. The horizontal axis is $X = \ln(\ln x)$ and increases up to about 15 million.

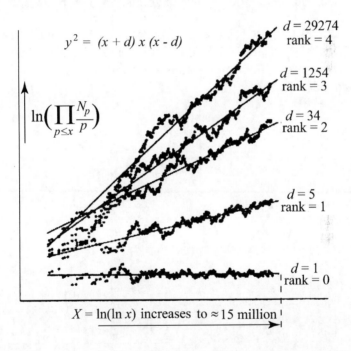

Figure 5.4. For easier visualization, this figure has been compressed vertically.

In the figure, the vertical axis has been squeezed down; in actuality, the rank 1 line has slope 1, with rank 2, 3, and 4 lines having slopes 2, 3, and 4. Also, the constants b for the rank 3 and 4 lines have been decreased, translating those two lines downward somewhat. These modifications are made to make the graph easier to fit on a piece of paper.

Figure 5.5 depicts the four elliptic curves $y^2 = (x+d)x(x-d)$ of rank $r = 0, 1, 2$, and 3. Once again, to make their plots fit nicely on the page, the x- and y-axes of curves of rank $r = 1, 2$, and 3 have been compressed.

Exercise 5.2.7. The abelian group of the curve $y^2 = (x+5)x(x-5)$ in Figure 5.5 is

$$\mathbb{Z} \oplus \mathbb{Z}_2 \oplus \mathbb{Z}_2,$$

and the figure indicates that $(-4, 6)$ generates $\mathbb{Z}$. Using your connect-the-dots code on the curve in $\mathbb{R}^2$, find the next few points in the real curve and then also in the curve modulo 29. (See Exercise 5.1.6 on p. 108.)

Exercise 5.2.8. In Figure 5.5, the drawings for $d = 5, 34$, and $1{,}254$ have been compressed vertically so that they look nearly identical to the curve for $d = 1$. For appropriate compressions and translations, can these all be made *exactly* identical?

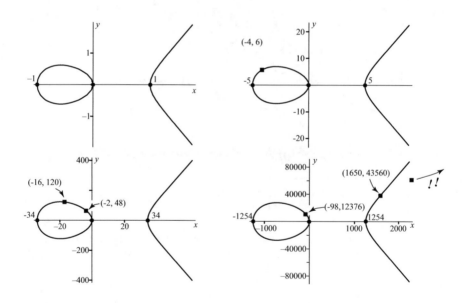

Figure 5.5. Here are the four elliptic curves $y^2 = (x+d)x(x-d)$ with $d = 1, 5, 34, 1{,}254$. Their respective ranks are $r = 0, 1, 2$, and 3.

5.3 The Notion of Expected Rank

In the next section we ask: If you take a random elliptic curve, what do you expect the rank to be? In some sense, we want to take fair, ever larger samples of elliptic curves, find the average rank of the curves in each sample, and then look for a trend as the sample size grows without bound. So we need to find some way of choosing fair samples.

Let's start with some notation. Because any rational elliptic curve has an equivalent representation as a curve having Weierstrass short form $y^2 = x^3 + ax + b$, one natural notation for such a curve is

Notation 5.3.1. Let $E(a, b)$ denote the rational elliptic curve defined by $y^2 = x^3 + ax + b$. ◊

From the boxed statement on p. 78, we can assume a and b are integers. Since $E(a, b)$ is to be nonsingular, the zeros of $y^2 = x^3 + ax + b$ must be distinct, meaning that the discriminant $\Delta = -(4a^3 + 27b^2)$ is nonzero. Statistically, it's relatively rare that (a, b) makes Δ zero — the point (a, b) would have to be on the cusp curve depicted in Figure 3.11 on p. 71. So it's safe to say that the integer-pairs (a, b) that are not on the curve $4a^3 = -27b^2$ essentially form the plane integer lattice $\mathbb{Z} \times \mathbb{Z}$, and we want to choose increasingly large samples from this lattice in a fair way. We do this in the next section. To get some perspective, let's look at some examples of fair sampling.

Example 5.3.2. In the world of all integers, if we want to see what proportion of integers are multiples of some fixed integer, a fair sample would be all the integers in a big interval of length N. As $N \to \infty$, the proportion of integers that are multiples of 10, say, would approach $\frac{1}{10}$. ◊

Example 5.3.3. From the Prime Number Theorem, we know that the intervals of integers $[0, \ldots, N]$ fairly sample the densities of primes as $N \to \infty$. The theorem says that the proportion of primes among the first N integers gets closer to $1/\ln(N)$ as N increases. This means, of course, that the densities of primes in these samples approach 0 as $N \to \infty$. ◊

Example 5.3.4. If we're sitting at the origin, then among integer lattice points (n_1, n_2) in the first quadrant, we'll be able to "see" a particular point (n_1, n_2) whenever n_1 and n_2 are relatively prime. In contrast, for any integer $|m| > 1$, (n_1, n_2) stands in the way of seeing (mn_1, mn_2). What are the chances that we'll be able to see any particular lattice point? If we

take our samples to be coordinate squares with diagonal points $(0,0)$ and (N,N), then the densities of lattice points visible from the origin approach $6/\pi^2 \approx 60.79\%$ as $N \to \infty$. For a proof, see [Apostol, Theorem 3.9, p. 63].
◇

5.4 Expected Rank of a Random Elliptic Curve

In this section we take a whole bunch of elliptic curves and ask, "What's the expected rank of a randomly chosen elliptic curve?" Do we expect it to have rank 0 on average? Rank 1 on average? Rank 20 on average? Does the average rank even have to be finite? After all, it's perfectly conceivable that there might be no upper limit to it — perhaps as we take larger and larger finite samples, the average grows without bound. It might happen, for example, that as we choose larger and larger coefficients in an elliptic curve's equation, the rank also grows in such a way that the average rank increases without any limit. Answers to such questions turn out to have a bearing on many questions in number theory. Of course, all such questions assume we have some way of choosing a "typical" or "random" elliptic curve. So we want to take samples that are both fair and as large as we want.

One way of doing this is to assign to each elliptic curve $E(a,b)$ a "size." We can then form samples of size bounded by a value such as a hundred or a thousand or a million, say. The term "height" captures this spirit and is a term routinely used in questions like this. Notice that Examples 5.3.2, 5.3.3, and 5.3.4 all use the phrase "as $N \to \infty$"; N is the size in each case.

As mentioned a moment ago, we can assume a and b in $E(a,b)$ are integers. As argued on p. 78, by appropriately stretching in the x- and y-directions by integer amounts, any $E(A,B)$ with rational A, B maps into an $E(a,b)$ with a,b integers. A straightforward argument shows that relevant facts about rational points on the elliptic curve are preserved under this map. Rational points map to rational points, nonrational points map to nonrational points, and the connect-the-dots algorithm works the same way, so the rank and torsion stay the same. Therefore instead of elliptic curves being tagged by points in $\mathbb{Q}^2$, we can tag them by points of the lattice $\mathbb{Z}^2$. Because of this, the spirit of Example 5.3.4 just above can be used in this definition of height of elliptic curves:

Definition 5.4.1. For a point (a,b) in the lattice $\mathbb{Z}^2$, the *height of* $E(a,b)$ is $h(E(a,b)) = \max(|a|^3, b^2)$. ◇

5.4. Expected Rank of a Random Elliptic Curve

Comment 5.4.2. The height basically measures the size of the coefficients a, b. A large height means that at least one of a, b is large, and largeness is correlated with greater arithmetic complexity of the curve. The powers of 3 and 2 in the definition serve as "scaling factors" to bring the height closer in spirit to the curve's discriminant Δ. The absolute value $|\Delta|$ itself is often used as the height, but for our purposes our definition turns out to be a bit simpler. Actually, many other notions of height appear in the literature, but these measures of size all tend to be about the same for all but a negligible proportion of elliptic curves. Here, for example, are three:

- The *naive height of* $E(a,b)$ is $\max(4|a|^3, 27b^2)$.
- The *Faltings height* is essentially the natural logarithm of the naive height.
- The *height of* $E(a,b)$ is often taken to be $|\Delta|$, the absolute value of the discriminant $\Delta = -(4a^3 + 27b^2)$. ◊

Now take a look at Figure 5.6. It's not only related to Example 5.3.4 on p. 120 in that we're choosing points in $\mathbb{Z}^2$, but in each case a "typical basket" is a rectangle. In Example 5.3.4 we looked at lattice points within increasingly larger squares. In Definition 5.4.1, it happens that the rectangle enclosing lattice points of heights $\leq N$ is defined in the real plane by $\max(|x|^3, y^2) = N$, which turns out to be the rectangle with corners $(\pm\sqrt[3]{N}, \pm\sqrt{N})$. As $N \to \infty$, the baskets form a sequence of increasingly large rectangles about the origin. Each lattice point in a basket of height N corresponds to an elliptic curve of height $\leq N$.

With this setup, we can now turn to finding the average rank of elliptic curves over larger and larger samples. Associated to each lattice point is an elliptic curve and therefore also a rank. For a height $N > 1$, write N^* for the number of lattice points in all four quadrants in the corresponding four-times-larger rectangle. Summing the ranks over all N^* points and dividing by N^* gives the average rank corresponding to height N. Of course finding the rank of each elliptic curve requires a fast computer, good theory, and a strategy for determining the rank, as well as good programming and a lot of patience. At this time, obtaining the average rank as $N \to \infty$ is a wish rather than an actual accomplishment.

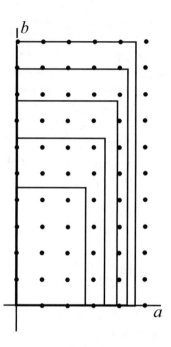

Figure 5.6. This picture shows only the first-quadrant snapshot of rectangles with corners ($\pm\sqrt[3]{N}, \pm\sqrt{N}$), together with the integer lattice points trapped within the rectangles. The picture in the other three quadrants is completed in the expected way. To keep the figure simple, the figure shows only the five rectangles corresponding to $N = 20, 40, 60, 80, 100$.

5.5 The Tale of Average Rank

Where are we in this journey? For many years, the belief was that half of all of elliptic curves have rank 0 and half have rank 1. That's not to say there aren't curves of rank 2 or higher. Computers have generated lots of them! But although there may be infinitely many curves of rank 2, 3, and higher, it was conjectured that those would be rare enough so that in large samples, half the curves would have rank 0 and half would be of rank 1. In fact, this conjecture was expected to hold no matter which one of the different notions of height was used in ordering the elliptic curves. However, computations on really huge samples of curves raised serious doubts about what people had believed for so long. Such computations were recently carried out by Bektemirov, Stein, and Watkins using

5.5. The Tale of Average Rank

the concept of "conductor," which is generally considered to be the most natural notion of height for elliptic curves over $\mathbb{Q}$. What they found is encapsulated in Figure 5.7. This is the graph of average rank, and it samples elliptic curves all the way up to height 10^8, which corresponds to around 18.5 million curves. Even from where the graph starts, the average rank is already higher than $\frac{1}{2}$, and the graph of average rank then seems to increase monotonically. Granted, it's not clear from this graph whether the average rank will continue increasing forever or whether it's actually approaching some asymptote. But from the data we have so far, it certainly doesn't appear to be approaching $\frac{1}{2}$.

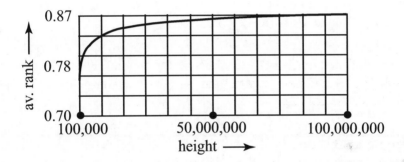

Figure 5.7. This graph was created by dividing the horizontal interval from a hundred thousand (10^5) to a hundred million (10^8) into many equal pieces. The conductor was used as a measure of height. Think of this plot as consisting of a succession of points, starting on the left with the average rank using a sample size of 100,000 curves, the next point being plotted for the average rank using 101,000 curves, then using 102,000 curves, and so on, up to the rightmost point corresponding to a sample size of 100,000,000 curves.

Comment 5.5.1. A few words about the "conductor," which is a basic positive integer associated with any elliptic curve: The number of elliptic curves cataloged in LMFDB is 3,064,705, and this is just the number of rational elliptic curves of conductor equal to or less than 500,000. ◇

However, based on theory and heuristics pointing to 50-50, Nicholas Katz of Princeton University and Peter Sarnak have for some time been confident that the graph will eventually turn around and come back to $\frac{1}{2}$. It's certainly fair to ask, "Is there any instance in mathematics where so many data points suggest one thing, but then still more data forces us to revise our guess?" The answer is yes. Here are two examples.

Example 5.5.2. One case in point is Polya's Conjecture which in one form says this: Suppose you color every integer $n > 1$ red if it has an odd number of (possibly repeated) prime factors, and green otherwise. Then no matter how large an N you pick, the integers less than N will always have more reds than greens. Computer tests show the conjecture to be true for every N up to 100 million, and in fact even up to 900 million. Then the conjecture breaks down at $N = 906{,}150{,}257$. (This was discovered in 1980 by Minoru Tanaka. See [Tanaka].) ◇

Sarnak is no stranger to questions of this general type — in 1994, he and Michael Rubinstein contributed to our understanding in this next example.

Example 5.5.3. The prime-counting function $\pi(x)$ expresses how many primes are equal to or less than the positive real number x. For many years it was conjectured that for all x, this step function π is less than its more nicely behaved estimate, the "logarithmic integral function" $\mathrm{li}(x) = \int_{t=0}^{x} \frac{dt}{\ln t}$. It has been shown that $\pi(x) < \mathrm{li}(x)$ for x up to at least 10^{19} — that is, 10 quintillion. Although as of this writing nobody has found a specific x for which the conjecture is false, it's been proved that there *does* exist such an x, and it is less than 1.4×10^{316}. (!) (See [Stoll].) ◇

So where do things stand? To begin, one can always compute average ranks over finite-sized samples of elliptic curves — that's due to Mordell's Theorem saying that the rank of any elliptic curve is finite. Figure 5.7 is an example of a big succession of ever-larger samples. Now let's extend the graph in this figure by allowing for a potential plot of average ranks over a far larger succession of samples — namely, intervals $[1, N]$, where N increases without bound. What happens to the average rank $\bar{r}_N$ as the average is computed over these successively longer intervals? Just looking

5.5. The Tale of Average Rank

at Figure 5.7, one might guess that $\bar{r}_N$ tends to something around 0.87 or 0.88. But depending upon the rank sizes encountered, our graph could continue rising or it might start to decrease or it could even keep wobbling around and never converge. Let's remember that the graph in Figure 5.7 is actually the plot of an initial part of the real-number sequence $\{\bar{r}_N\}$, but where the horizontal axis of integers has been so greatly compressed that the plot looks like the graph of an ordinary function.

We can't say whether or not $\{\bar{r}_N\}$ converges, but we can construct a related sequence that does. We do this as follows: First, it's known that $\{\bar{r}_N\}$ is bounded (see [Bhargava 6]), so $\{\bar{r}_N\}$ has a least upper bound. Second, for each n, consider the "tail" of $\{\bar{r}_N\}$ consisting of all those $\bar{r}_N$ with $N \geq n$. With increasing n, we get a decreasing chain of nested subsets, and that means each such tail is bounded. Third, the sequence $\{r_n\}$ of least upper bounds of these ever-shorter tails can never increase with n, so $\{r_n\}$ monotonically decreases and is bounded from below since no rank is negative. Finally, the above means that $\{r_n\}$ must converge to some real value $\mathbf{r}$. (This is called the limit superior or limsup of $\{r_n\}$ and is the largest possible limit of any convergent subsequence of $\{r_n\}$.) The very big question is:

> What's the value of $\mathbf{r}$?

The answer is that we don't know. But as time progresses, so do we. There have been some notable advances through the years, but each proof assumes both the Generalized Riemann Hypothesis (GRH) and the Birch and Swinnerton-Dyer Conjecture (BSD) — in other words, each advance assumes two million dollars worth of Clay Institute conjectures! (Today, GRH and BSD are two of the most important conjectures in mathematics; as of this writing, neither has been proved.) Here are some of these advances:

- 1992: Armand Brumer of Fordham University proved that $\mathbf{r} \leq 2.3$.
- 2004: Roger Heath-Brown improved Brumer's result to $\mathbf{r} \leq 2$.
- 2007–2009: Bektemirov, Stein, and Watkins carried out extensive computations leading to a graph as in Figure 5.7 on p. 123 suggesting (but not proving) an upper bound of around 0.88.
- 2009: Matthew Young was able to move Heath-Brown's result down to $\mathbf{r} \leq 25/14$. Since 25/14 is strictly less than 2, this means that assuming GRH and BSD, the curves having rank 0 or 1 make up a positive proportion of rational elliptic curves.

- 2016: A computational breakthrough was achieved by Balakrishnan, Ho, Kaplan, Spicer, Stein, and Weigandt; the paper appears in [BHKSSW]. This is of great importance because the plot of hard data depicted in Figure 5.8 shows the graph turning around and decreasing. Using a more common notion of height than conductor, they found that the graph of average rank actually reaches some value greater than 0.89 but then decreases to an upper bound of 0.885.

- Katz and Sarnak based their prediction on very good random matrix models and various other kinds of heuristics strongly suggesting 0.5. If their prediction of **r** = 0.5 is true, intervals longer than 27 billion will be needed to see this as a data-driven plot.

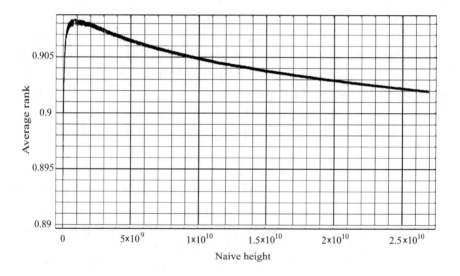

Figure 5.8. This is the dramatic graph appearing in [BHKSSW], showing that the graph on p. 123 does eventually turn around! The graph here goes out 270 times farther, to 27 *billion*.

- There's now an additional twist to the story. Bhargava and his colleague Arul Shankar have recently given a very different argument, based on the "4-Selmer group" which likewise says that the graph must not only turn around, but approach 0.5. So although as of this writing there's no proof, all these researchers currently share the same opinion: Given a large enough sample size of elliptic curves over $\mathbb{Q}$, the average size is just what people believed is was years ago, 0.5.

Exercise 5.5.4. In an $N \times N$ square of integer lattice points, every lattice point is painted green, while in an identical square, the lattice points off either diagonal are colored red. Certainly the proportion of green points in the first square is 100%. Prove whether or not the proportion of red points in the other square approaches 100% as $N \to \infty$. Plot the ratios of greens to reds as $N \geq 3$ increases without bound.

5.6 Rank Results Without GRH or BSD

As we mentioned earlier, two of the most important conjectures in mathematics are the Generalized Riemann Hypothesis (GRH) and the Birch and Swinnerton-Dyer Conjecture (BSD). Every one of the above results assumes *both* of these mega-conjectures! Can anything be said without making these major assumptions? We stated on p. 93:

> For the vast majority of questions about the rank of elliptic curves, the answer is unknown. There are lots of reasonable conjectures and lots of data, but few proofs.

Manjul Bhargava of Princeton together with his former graduate student Arul Shankar have put a dent in this statement, supplying proofs of some basic conjectures and turning them into important theorems. Let's briefly describe them.

Theorem 5.6.1.

> If elliptic curves are ordered by height, then their average rank is less than 1.

This is a breakthrough, because we now definitely know that rank 0 curves represent a positive proportion of all elliptic curves. Everyone believed it, but now, finally, there's proof. What does "positive" mean? Did they prove that only a millionth of all curves have rank 0? Were they able to in fact guarantee a more substantial share? They established

Theorem 5.6.2.

> At least 10% of all elliptic curves are of rank 0.

And what about the larger set of curves of rank 0 or 1? Here again, Bhargava and Shankar made a significant breakthrough. For years, it was believed that curves of rank 0 or 1 so outnumbered all others, that their proportion in the universe of all elliptic curves was 100%. Data suggested this, but nobody knew for sure. They dramatically changed the landscape by proving

Theorem 5.6.3.

> The proportion of curves having rank 0 or 1 is at least 80%.

The goal is still 100%, but they have put a big hole in the armor of what seemed to be an impenetrable problem.

They also made progress on the Birch and Swinnerton-Dyer Conjecture. It had previously been proved in many cases of rank 0 and 1 due to the work of Gross-Zagier and Kolyvagin, but in the 2014 paper [BSZ], Bhargava, Christopher Skinner, and Wei Zhang have pushed the envelope by proving that these methods cover a very large swath of elliptic curves:

Theorem 5.6.4.

> The Birch and Swinnerton-Dyer Conjecture is true for at least 66.48% of all elliptic curves.

It is fair to ask how they went about proving these theorems. This work is highly technical and requires specialized background in algebraic number theory, analytic number theory, and algebraic geometry. Fortunately, Bhargava has endeavored to communicate those methods to a wider audience. Here, in his own words, is a simplified description.[1]

---

[1] Quotation taken from MAA transcription of Manjul Bhargava's Power-Point Hedrick Lectures in the summer of 2011 in Lexington, Kentucky. © Mathematical Association of America, 2021. All rights reserved.

5.6. Rank Results Without GRH or BSD

"To get started, my graduate student Arul Shankar and I actually went back to one of the old computer algorithms of BSD, which are full of cleverness, and we thought, 'Let's use some of their cleverness to a theoretical end.' We used some of their algorithms and tried to put them in a theoretical framework to see what the algorithms that led to their conjecture would say about average rank.

"What the work boils down to is this: Shankar and I found a way of injecting rational points on elliptic curves into lattice points in $\mathbb{R}^n$, turning the question of proving that the average rank is finite into computing the number of lattice points in regions within $\mathbb{R}^n$. Normally, one expects that the number of lattice points in a region of $\mathbb{R}^n$ should be about the volume of that region. But in our case, the regions that naturally arise in these kinds of hard number theory problems are never compact. In fact, they can have tentacles that can go out to infinity, and for any one of them, you never know whether it's passing through infinitely many lattice points or whether it's totally avoiding all lattice points. So a lot of our work went into developing techniques to count lattice points in thin regions that go off to infinity.

"That's what much of our work is about, and a lot of analytic number theory is involved to see if those lattice points that you really are interested in are the images of rational points on elliptic curves. This was something we worked on very hard for a year. In the end, it was successful, and we also found some extensions of the original BSD techniques to various other scenarios. When we put all that together, we actually ended up counting points in fifty-dimensional spaces.

"As for the Birch and Swinnerton-Dyer Conjecture, we got our results by taking the original arguments Birch and Swinnerton-Dyer used, but turned them backwards, then found a new technique to avoid the Birch and Swinnerton-Dyer Conjecture. We used that to try to prove many special cases of the Birch and Swinnerton-Dyer Conjecture."

5.7 About Manjul Bhargava

Manjul Bhargava was born in 1974 in Canada, but his family soon settled in Long Island, New York. His mother, Mira Bhargava, an Emeritus Professor of mathematics at Hofstra University, was his first mathematics teacher. From early childhood, Manjul was strongly attracted to mathematics. His mother recalls that the only way she could get her energetic 3-year-old to stay still was to give him big numbers to add or multiply. "Instead of using paper and pencil, he would kind of flip his fingers back and forth and then give me the right answer. I always wondered how he did it, but he wouldn't tell me. Perhaps it was too intuitive to explain."

Figure 5.9. Manjul Bhargava[2]

Around that time, Manjul had already revealed a burgeoning interest in musical rhythms and asked his mother to teach him to play the tabla (consisting of a pair of hand/finger drums). He responded intensely to sequences of long and short notes, which has inherent in it an element of combinatorics. For example, if "1" represents a quarter note, "2", a half note, "3" a dotted half, and "4", a whole note, then there are eight possible rhythmic patterns within the duration of a whole note: 1111, 211, 121, 112, 22, 31, 13, and 4. Listening to, playing, and absorbing these rhythms grew to be part of his inner nature, and today, when he gets stuck on a mathematics problem, he will often turn to the tabla. After a good session, he finds his mind has cleared.

As luck would have it, Manjul's maternal grandfather was a Sanskrit scholar and the youngster was introduced to Sanskrit poetry. In the same way that Manjul was captivated by intricate rhythms in Indian music, he was taken by Sanskrit poetry — it similarly has rhythms arising from long

---

[2]Photo source: Wikimedia Commons, author: IMU, used under the Free Art License. http://artlibre.org/licence/lal/en/.

5.7. About Manjul Bhargava

and short syllables. Bhargava himself put it this way: "Mathematics, music, and poetry together feel like a complete experience. All kinds of creative thoughts come together when I think about all three."

Rhythms, based on flowing time, are essentially 1-dimensional. By the time he was eight, numbers took on a higher-dimensional life for him as well. Before a basket of oranges would go into the family juicer, Manjul would stack them into a triangular pyramid, and he wondered if there was an easy way to find the number of oranges without actually having to count them. He became really fascinated with this problem and turned it over in his mind for a long time. Finally, after several months, he saw that if an edge of a triangular pyramid is n oranges long, then the total number of oranges in the standard, closest packed pyramidal arrangement is

$$\frac{n(n+1)(n+2)}{6}.$$

That was a tremendously exciting moment and gave him a real taste for the predictive power of mathematics.

Given his talents, it's hardly surprising that Manjul disliked school. He'd frequently ask his mother if he could go to work with her instead of having to suffer through yet another day of grade school. She was understanding, and the 8-year-old often skipped school, instead exploring Hofstra's library, walking around in the arboretum, and sometimes attending an elementary probability class she was teaching. After regular school let out, however, Manjul could often be seen playing basketball or tennis with his classmates. He always wore his talents lightly and was known as a "regular, friendly guy." Over the years, Manjul missed a lot of school, one contributing factor being trips to India. He loved those trips. He was just as friendly with animals as with humans, and he tells the story about one visit to India. "One day I saw three peacocks — one adult and two children — walking around in our garden, so I thought it might be nice to feed them something. I invited them inside the veranda, found the family's store of 'best-grade grains,' and the peacocks eagerly gobbled up the unexpected offering. The very next morning I heard a pecking on the door, and to my surprise, the same three peacocks had come to visit again for some breakfast. Of course they got what they wanted, and this immediately established a morning ritual of pecking and feeding. Eventually, the family discovered what had happened to their store of grains and the peacocks had to deal with disappointment."

Chapter 5. A Million-Dollar Challenge

All through grade school, Bhargava was undecided about what he would do in life. He had already become virtually professional on the tabla, making music a very real possibility. His intense encounter with Sanskrit poetry made him think seriously about linguistics as a career. He also considered being an economist and even a mountain climber. He eventually whittled it down to mathematics versus music. (Many mathematicians know the feeling.) Finally, once at Harvard, he decided on math, realizing that it was always the mathematical aspects of these choices that got him most excited. "Somehow, I always came back to math."

At Harvard he was an undergraduate teaching assistant. His simple approach together with his friendly and unassuming personality made him a popular and effective teacher, and for three consecutive years he won the Derek C. Bok Award for teaching excellence. After graduating from Harvard, he went to Princeton for graduate study, where he focused intensely on his passion, number theory.

Even from the days of Diophantus and later Fermat, a central question in number theory had to do with squares or sums of squares. For example, Diophantus noted that no number of the form $4n + 3$ (or equivalently, $4n - 1$) can be the sum of two squares. Later, Fermat wrote that every natural number can be expressed as the sum of at most four squares and that every prime of the form $4n + 1$ is the sum of two squares in exactly one way, but that every prime of the form $4n - 1$ is *never* the sum of two squares. Fermat also proved that the area of any Pythagorean triangle (a right triangle with integer sides) is never a square. A monumental step forward was made by Karl Gauss in his *Disquisitiones* where he considered the more general question of binary forms $ax^2 + 2bxy + cy^2$ with positive integer coefficients. Gauss found an ingenious way of composing two binary forms to create a third. As a very special example of this, he showed that *if two integers are each the sum of two squares, then so is their product*.

Example 5.7.1. Let's illustrate the statement above, "If two integers are each the sum of two squares, then so is their product." Take $2^2 + 3^2 = 4 + 9 = 13$ as one integer and $4^2 + 5^2 = 16 + 25 = 41$ as another. Their product 13×41 is 533, and this is the sum $7^2 + 22^2 = 49 + 484 = 533$. As a somewhat wilder example, the product of the two integers $5^2 + 11^2 = 146$ and $17^2 + 23^2 = 818$ is $146 \times 818 = 119{,}428$ which is the sum of two squares in two different ways: $338^2 + 72^2$ and $302^2 + 168^2$. For other choices,

5.7. About Manjul Bhargava

there can be even more ways: For example, the two integer right triangles $(3, 4, 5)$ and $(5, 12, 13)$ tell us that 5^2 is the sum $3^2 + 4^2$ and $13^2 = 5^2 + 12^2$. Therefore 5^2 and 13^2, each being the sum of two squares, means their product $5^2 \times 13^2 = 25 \times 169 = 4{,}225$ must be the sum of two squares, and this can happen in *five* different ways:

$$0^2 + 65^2 = 4{,}225,$$
$$16^2 + 63^2 = 256 + 3{,}969 = 4{,}225,$$
$$25^2 + 60^2 = 625 + 3{,}600 = 4{,}225,$$
$$33^2 + 56^2 = 1{,}089 + 3{,}136 = 4{,}225,$$
$$39^2 + 52^2 = 1{,}521 + 2{,}704 = 4{,}225. \quad \diamond$$

Gauss takes around 20 pages for his treatment of combining quadratic forms. Bhargava read through it all and left thinking that there must be a better way to do this. The *Disquisitiones* was then already 200 years old, and nobody had found any better way. In fact, Gauss's outcome was considered by many to be something of a curiosity, an isolated result.

Bhargava was meditating about this problem one night in California in a room strewn with a variety of puzzles, including a Rubik's "minicube." Instead of being a $3 \times 3 \times 3$ arrangement of 27 cubes, the minicube is a $2 \times 2 \times 2$ arrangement of 8 cubes. He thought of labeling each of the outermost 8 corners with a number. The six faces of the cube give rise to three pairs of opposite faces. Each (square) face has four vertices with each vertex having its associated number, and these four vertices form in a natural way a 2×2 matrix. There is thus a pair of 2×2 matrices corresponding to any pair of opposite faces. Denoting a typical such pair of matrices by A and B, we can form the determinant $\det(Ax - By)$, where x and y are indeterminates. If we write

$$A = \begin{pmatrix} a_{11} & a_{12} \\ a_{21} & a_{22} \end{pmatrix} \quad \text{and} \quad B = \begin{pmatrix} b_{11} & b_{12} \\ b_{21} & b_{22} \end{pmatrix},$$

then $\det(Ax - By)$ becomes

$$\det \begin{pmatrix} a_{11}x - b_{11}y & a_{12}x - b_{12}y \\ a_{21}x - b_{21}y & a_{22}x - b_{22}y \end{pmatrix}$$
$$= (a_{11}x - b_{11}y)(a_{22}x - b_{22}y) - (a_{12}x - b_{12}y)(a_{21}x - b_{21}y),$$

a quadratic form in x and y. There's a similar quadratic form in x and y for each of the other two pairs of cube-faces, and any two of them can be looked at as the input to Gauss's composition rule, and the third, as the output. Bhargava realized he had found a neat, geometric reformulation of Gauss's 20-page work. Bhargava recalled, "That one night was one of the most exciting in my life!" Over the next few years, he found another 13 ways of composing quadratic and higher-degree forms — ways Gauss never thought of — and ushered in a whole new era of studying such forms. Besides the remarkable progress he made on ranks of elliptic curves and the Birch–Swinnerton-Dyer Conjecture, the Rubik's Cube-inspired discovery figured into awarding Bhargava a 2014 Fields Medal.

No biographical sketch of Bhargava is complete without at least mentioning two other results, "The 290 Theorem" and "The 15 Theorem." Based on examples from *Arithmetica*, it seems almost certain that Diophantus was aware of what we call "The Four Squares Theorem": Every positive integer is the sum of four squares $x^2 + y^2 + z^2 + u^2$, where x, y, z, u are integers. Fermat stated and claimed to have proved it, but it was Joseph Lagrange who in 1770 finally definitely established it. The quadratic form $x^2 + y^2 + z^2 + u^2$ is called *universal* because every integer can be expressed in this way. There are 53 other "pure" universal forms $ax^2 + by^2 + cz^2 + du^2$ (a, b, c, d integers), and Ramanujan found them all.

However, what about all the other quadratic forms — those with quadratic mixed terms like xy, xz, and so on? There are many more such universal forms that, like $x^2 + y^2 + z^2 + u^2$, are positive for any choice of the integer variables except when all four variables are zero. Based on both theory and computer data gathered by two of his students, John H. Conway in 1993 conjectured that such a form is universal if and only if you can find instances of integers x, y, z, u for which the form assumes each of the 29 values

$$1, 2, 3, 5, 6, 7, 10, 13, 14, 15, 17, 19, 21, 22, 23, 26, 29,$$
$$30, 31, 34, 35, 37, 42, 58, 93, 110, 145, 203, 290.$$

In this list of 29 numbers, the largest one is 290, giving the theorem the nickname "The 290 Theorem." Bhargava and his colleague Jonathan Hanke proved this remarkable theorem in 2005. They also established the precise number of such universal quadratic forms in four variables — it's 6,436.

5.7. About Manjul Bhargava

The 290 Theorem has a little brother, "The 15 Theorem." Here, the integer coefficients of the mixed terms are all even. The spirit is that since multiplication among integers is commutative, a mixed term like yz is the same as zy. If each occurs with an integer coefficient, then the total for both is double an integer — an even one. With this restriction, there are fewer forms and fewer cases to test. Conway had not only conjectured but also proved with his student William Schneeberger that a form like this is universal if you can find integers x, y, z, u making the form have each of the nine values

$$1, 2, 3, 5, 6, 7, 10, 14, 15.$$

His proof was quite complicated and was never published. Bhargava gave a simpler and more elegant proof in 2000.

Peter Sarnak of Princeton University and the Institute for Advanced Study once said about Bhargava, "Anything he touches, glows. After he's thought about it . . . the field looks very different." In a similar vein, Andrew Granville at the University of Montreal summed up Bhargava's remarkable mathematical intuition this way: "He has his own perspective that is remarkably simple compared to others. Somehow, he extracts ideas that are completely new or are retwisted in a way that changes everything. But it all feels very natural and unforced — it's as if he found the right way to think."

6

Every Real Elliptic Curve Lives in a Donut

6.1 Complex Curves

Elliptic curves defined over $\mathbb{Q}$ — the rational elliptic curves — are standouts because they are so helpful in understanding and solving homogeneous third-degree number theory problems. But elliptic curves can just as well be defined over $\mathbb{R}$ or $\mathbb{C}$, and in these settings they reveal markedly different personalities. Studying them leads to surprises and deep connections. We now begin this journey. Here, the trio of fields

$$\mathbb{Q} \subset \mathbb{R} \subset \mathbb{C}$$

is an overall organizing principle that's as powerful as it is simple.

In this book we've so far worked mainly at the $\mathbb{Q}$ level, converting rational homogeneous number theory problems of degree at most 3 into lines, nondegenerate conics, and cubics. Geometry then helps us arrive at solutions. In $\mathbb{Q}^2$ we have seen that when they exist, rational points are everywhere dense in a nondegenerate conic, while for a cubic, in many cases there are only finitely many rational points. At the other extreme of $\mathbb{C}$, these degree-one, -two, and -three curves, respectively, become, topologically, a sphere, a sphere, and a torus. The rational points lying within any one of these surfaces constitute a very skinny subset and, amazingly, the genus of the full surfaces — a very global concept which Chapter 7 covers in more detail — tells us quite a bit about the nature of that skinny set of rational points. This remarkable linkage continues to hold even in higher degrees, where the genus is larger and number theory solutions are always sparse.

6.2 Complex Numbers Enlighten

The term "complex setting" can be taken in two different ways. One is in the affine sense as a direct analog of $\mathbb{R}^2$, namely $\mathbb{C}^2 = \mathbb{C} \times \mathbb{C}$. It can also be taken in the projective sense as a direct analog of $\mathbb{P}^2(\mathbb{R})$. In the first sense, $\mathbb{C}^2$ is just a vector space of complex dimension 2 — that is, the set of all ordered pairs (x, y) with x and y in $\mathbb{C}$ and supplied with scalar multiplication by elements of $\mathbb{C}$ as well as vector addition. As for the complex analog of $\mathbb{P}^2(\mathbb{R})$, we can proceed in essentially the same way as in the real case, which we looked at as the unit disk with antipodal points identified. This is depicted in Figure 2.2 on p. 40, where the map $r \to \frac{r}{|r|+1}$ compresses $\mathbb{R}$ to the open interval $(-1, 1)$, which is a "unit open 1-disk" — that is, an open 1-dimensional disk or interval of radius 1 (that is, diameter, or length, 2). We can do an analogous thing in the complex setting. If $|c|$ denotes the real distance from the origin to $c \in \mathbb{C}$, then the map $c \to \frac{c}{|c|+1}$ compresses $\mathbb{C}$ to the open 2-disk $|c| < 1$. Continuing our analogy to the real case, perform this shrinking on each complex 1-subspace of $\mathbb{C}^2$, the points of a complex 1-subspace being all $\mathbb{C}$-scalar multiples of some fixed nonorigin point in $\mathbb{C}^2$. The union of all these shrunken complex 1-subspaces is a subset of $\mathbb{C}^2$ and is analogous to the open real disk in $\mathbb{R}^2$. Now in the real case, we added boundary points of each 1-disk and identified antipodal points. That's equivalent to looking at any shrunken real line — an open interval bridging two endpoints — as a circle with a missing point, and adding that point to make a closed loop. The complex analog regards a shrunken $\mathbb{C}$ as a sphere with a missing point — essentially a Riemann sphere without its north pole. Now add that missing point. This collection of Riemann spheres is analogous to the set of circles in $\mathbb{P}^2(\mathbb{R})$. (Those nine lines in Figure 2.3 on p. 41, with antipodal points identified, are topologically nine circles.) We denote this complex analog as $\mathbb{P}^2(\mathbb{C})$. This is difficult to visualize, but we can informally think of $\mathbb{P}^2(\mathbb{C})$ as $\mathbb{C}^2$ to which we've added points at infinity. This will become clearer as we consider some examples.

6.3. Plotting a Complex Circle

Exercise 6.2.1. When the elliptic curve $y^2 = x^3 - x - 1$ is plotted in $\mathbb{C}^2$, most of its points are complex. $P = (0, i)$ is an example. The pseudocode in Appendix C styled for use by Maple or Mathematica can compute successive multiples of P and is algebraic in nature, so code working in the real rational setting can also calculate successive multiples of any complex P on the curve. Find $2P$, $3P$, and

$$4P = \left(\frac{223}{784}, \frac{24{,}655 i}{21{,}952}\right).$$

Check that each of your points does in fact satisfy $y^2 = x^3 - x - 1$.

6.3 Plotting a Complex Circle

Example 6.3.1. Plotting a circle in the complex setting is illuminating and sets the stage for analogous plots of elliptic curves. Let's see what happens in the case of, say, the circle $x^2 + y^2 = 1$. We now assume x and y are complex, so we accordingly write $x = x_1 + ix_2$ and $y = y_1 + iy_2$. Substituting these into $x^2 + y^2 = 1$ and separating into real and imaginary parts gives

$$x_1^2 - x_2^2 + y_1^2 - y_2^2 = 1,$$

$$x_1 x_2 + y_1 y_2 = 0.$$

It turns out that each of these equations defines a real 3-dimensional surface in $\mathbb{R}^4$ and that their intersection is a real 2-dimensional surface in $\mathbb{R}^4$. Of course it's hard for most of us to visualize in 4-space, but one thing we can do is cut everything down by one dimension. This can be done by setting $x_2 = 0$, which defines a real 3D slice in $\mathbb{R}^4$. This 3D slice — (x_1, y_1, y_2)-space — intersects the real 3-dimensional surface in a real curve. By setting $x_2 = 0$, only x_1, y_1, and y_2 can assume nonzero values. With $x_2 = 0$, the two equations above reduce to

$$x_1^2 + y_1^2 - y_2^2 = 1,$$

$$y_1 y_2 = 0.$$

Now $x_1^2 + y_1^2 - y_2^2 = 1$ defines a hyperboloid of one sheet, and $y_1 y_2 = 0$ defines the union of the (x_1, y_2)-plane when $y_1 = 0$, together with the (x_1, y_1)-plane when $y_2 = 0$. The two equations mean that we don't see the whole hyperboloid, but only the part within these two planes. Figure 6.1 depicts this curve in (x_1, y_1, y_2)-space. ◊

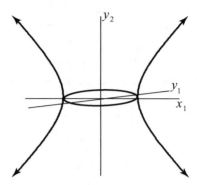

Figure 6.1. The intersection in $\mathbb{R}^4$ of the complex unit circle with (x_1, y_1, y_2)-space.

One of the things we've learned from the projective disk model $\mathbb{P}^2(\mathbb{R})$ is how points at infinity glue together the "ends" of branches to form topological loops. It turns out that the same thing happens in our complex analog, $\mathbb{P}^2(\mathbb{C})$. The hyperbola in the (x_1, y_2)-plane has two asymptotes, and at the "ends" of each asymptote, branches get glued together. For example, to go from Figure 6.1 to Figure 6.2, think of the hyperbola in Figure 6.1 as made of wire. Bend the two topmost wires in the (x_1, y_2)-plane — always keeping them in that plane — so that the two ends almost meet to make the top half of a circle. Repeat, using the bottommost two wires to make the bottom half of a circle. Then, just before soldering the ends together, twist the right half of the circle 180 degrees about the x_1-axis, and then do the soldering. The two soldered joints are the two points "∞" in the figure. Topologically, we get one big loop having two solder joints, with this big loop still touching the real circle at 1 and -1 in Figure 6.2.

Now our space $\mathbb{R}^3$ defined by $x_2 = 0$ is but one 3-dimensional slice in $\mathbb{R}^4$. More generally, any $x_2 = r = $ a constant defines a 3-space parallel to the slice $x_2 = 0$. As r varies through $\mathbb{R}$, these slices fill 4-space. As an example, let's vary x_2 a little — say, to $x_2 = r = 0.1$. With $x_2 = 0.1$, $x_1^2 - x_2^2 + y_1^2 - y_2^2 = 1$ then becomes $x_1^2 + y_1^2 - y_2^2 = 1.01$. This still defines a hyperboloid of one sheet, but instead of being intersected by the two planes defined by $y_1 y_2 = 0$, the intersection is with $y_1 y_2 + 0.1 x_1 = 0$, another hyperboloid of one sheet, but which looks quite a bit like two planes. This is depicted as the shaded surface in Figure 6.3, and the

6.3. Plotting a Complex Circle

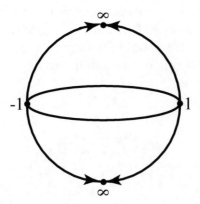

Figure 6.2. Here, the skeleton curve in Figure 6.1 has been topologically massaged to fit on a sphere.

previous circle-plus-hyperbola has morphed slightly into two curves, each running close to the circle-plus-hyperbola. For clarity, we have drawn just one of these.

If we topologically transform Figure 6.3 in the same way that Figure 6.1 got transformed into Figure 6.2, our curve now appears somewhat

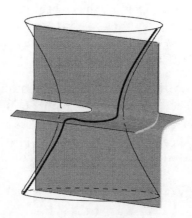

Figure 6.3. This shows one of the two disjoint curves comprising the intersection of the complex circle with the slice $x_2 = r = 0.1$.

like the largest loop in the top half of Figure 6.4, and the other curve (the one we didn't draw in Figure 6.3) appears dashed on the rear lower half. We see other loops corresponding to other choices of $r > 0$, and these loops fill out half the sphere, the other half being filled out as r runs through negative values. ◇

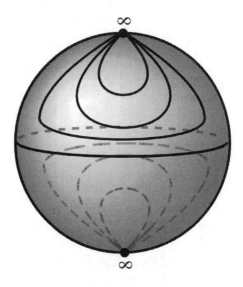

Figure 6.4. As r in Figure 6.3 takes on all real values, the sphere gets covered with disjoint curves. A few are shown here.

Exercise 6.3.2. Follow the approach used for plotting a complex circle to plot the (x_1, y_1, y_2)-slice of the complex parabola $y = x^2$. Topologically map this skeleton to a sphere. Then plot the slices $x_2 = r$ (r small) to see how nearby parallel slices map to the sphere. Finally, indicate how the remaining curves fit on the sphere by sketching in a few more curves.

Exercise 6.3.3. Redo Exercise 6.3.2 for the complex hyperbola $xy = 1$.

6.4 Plotting a Complex Elliptic Curve

Let's now explore plotting an elliptic curve in the complex setting. We can mimic the steps used for a unit circle in the last section.

6.4. Plotting a Complex Elliptic Curve

Example 6.4.1. For simplicity, we choose the elliptic curve $y^2 = x^3 - x$, shown in Figure 6.5, where we see both its affine plot and its image in the projective disk. As with the circle, assume x and y are complex and set $x = x_1 + ix_2$ and $y = y_1 + iy_2$. Substituting these into $y^2 = x^3 - x$ and separating into real and imaginary parts gives

$$y_1^2 - y_2^2 = x_1^3 - 3x_1 x_2^2 - x_1,$$
$$2y_1 y_2 = 3x_1^2 x_2 - x_2^3 - x_2.$$

Again, as with the circle, we first look in the 3-space $x_2 = 0$, and in that (x_1, y_1, y_2)-space the two equations lead to

$$y_1^2 - y_2^2 = x_1^3 - x_1 \text{ and } y_1 y_2 = 0.$$

As before, $y_1 y_2 = 0$ defines the union of the (x_1, y_1)- and (x_1, y_2)-planes. In the (x_1, y_1)-plane defined by $y_2 = 0$, the equation $y_1^2 - y_2^2 = x_1^3 - x_1$ becomes $y_1^2 = x_1^3 - x_1$ and we see the real elliptic curve we started with. In the (x_1, y_2)-plane given by $y_1 = 0$, the equation $y_1^2 - y_2^2 = x_1^3 - x_1$ becomes $-y_2^2 = x_1^3 - x_1$, and we see essentially the reflection of the curve about the y_1-axis but drawn in the (x_1, y_2)-plane. These curves are depicted in Figure 6.6, where the more lightly drawn part is the reflected curve.

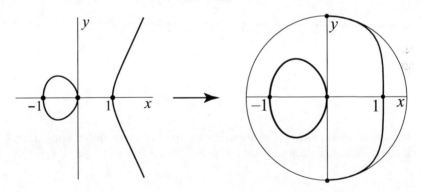

Figure 6.5. Image in the projective plane of a cubic.

In this figure, we see two branches. The one in the (x_1, y_1)-plane joins ends at the end of the y_1-axis, and the other joins its ends at the end of the y_2-axis. Since we're in the complex setting, the (y_1, y_2)-plane is the Riemann sphere minus its point at infinity, and the two branches both join up at this one point at infinity, $\{\infty\}$. So topologically, one loop goes from $x_1 = 1$ to $\{\infty\}$, with another one going from $x_1 = -1$ to $\{\infty\}$. Together with

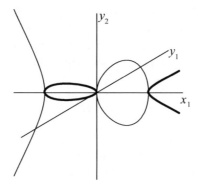

Figure 6.6. The heavily drawn curves are what we see in the usual (x_1, y_1)-plane. Looking in the (x_1, y_2)-plane reveals a "reflected" image of this.

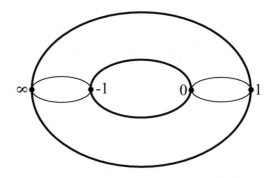

Figure 6.7. The ordinary and reflected curves in Figure 6.6 topologically lie on a torus in a natural way.

the other two loops touching the origin, there are four loops. In Figure 6.6, two of the loops can be regarded as made of thin wire, and the other two, out of thick wire. These wires can be bent to form the skeleton depicted in Figure 6.7. In analogy to the circle example, we can now let x_2 run through real values r, and the corresponding curves fill out the torus. The curves for $r > 0$ are depicted in Figure 6.8 and fill out half the torus — the upper-front and lower-rear quarters in the picture. The lower front and upper rear are filled in as r runs through negative values. ◊

6.5. Subgroups and Cosets

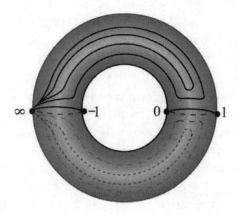

Figure 6.8. The curves corresponding to setting x_2 equal to real constants topologically cover the torus with curves, any two touching only at the point at infinity.

Exercise 6.4.2. For the complex curve $y^2 = x(x^2 - 1)(x^2 - 4)$, use the methods of this section to topologically map intersections of the curve with 3D slices of $\mathbb{R}^4$ to a double torus like the one in Figure 7.1 on p. 156.

6.5 Subgroups and Cosets

When we plotted an elliptic curve in the complex setting, curves over $\mathbb{R}$ played a pivotal role since the complex curve was built up as the union of real curves. In fact, the very suggestive torus skeleton depicted in Figure 6.7 is topologically the union of, for example, the real elliptic curve $y^2 = x^3 - x$ and its "brother" $y^2 = -x^3 + x$, obtained by replacing x everywhere by $-x$. There are two groups within this skeleton, one of them consisting of the heavily drawn curves in Figure 6.7 and the other one comprised of the lightly drawn curves. It's not hard to see this fundamental fact shared by elliptic curves over $\mathbb{Q}$, $\mathbb{R}$, and $\mathbb{C}$:

> The connect-the-dots addition algorithm works in all three settings. Therefore an elliptic curve in $\mathbb{P}^2(\mathbb{C})$ is an abelian group, as are the real and the rational subsets of the curve.

> From now on, the term *group* will always mean abelian group.

To keep things simple, we continue to work with the familiar example of $y^2 = x^3 - x$ shown in the left picture of Figure 6.5, although our arguments are basically the same for any short-form elliptic curve with an oval. (If there's no oval in the usual real plane, it gets revealed by looking in a different plane, as we will see in the next section.) The curve $y^2 = x^3 - x$ is an abelian group in $\mathbb{P}^2(\mathbb{R})$ under our connect-the-dots algorithm for which a line cutting the curve is basic. From just looking at the curve, it seems that any line cuts the real curve in exactly one or three points. This is easy to check. On the one hand, when $m \neq 0$, we can substitute $mx + b$ for y in $y^2 - x^3 + x = 0$ and this results in a cubic in x which factors into three linear terms. If their roots are all real, the line cuts the curve in three real points. If the roots are not all real, then because complex roots come in conjugate pairs, there's just one real root and the line cuts the real curve in a single point. On the other hand, if the line is vertical, say $x = x_0$, then plugging x_0 in for x in $y^2 - x^3 + x = 0$ gives y^2 equal to some constant $k = x_0^3 - x_0$. There are three cases. If $k > 0$, then y has two distinct solutions which, with the point at infinity, give three points of intersection. If $k = 0$, then y^2 has two identical solutions, meaning the line $x = x_0$ is tangent to the oval and this double point plus the point at infinity again gives three intersections. $k < 0$ means the only point of intersection is the point at infinity.

This leads to two observations about the connect-the-dots algorithm.

- The first is that any line intersects the branch in at least one point. If the line is vertical, it intersects the branch at infinity. If it's not vertical, the reader can pin this case down in Exercise 6.5.8 on p. 148.

- The second observation is that any line intersects the oval in either zero or two points. To see this, suppose a line intersects the oval in a point P. As a second point Q travels once around the oval (starting, say, from P as a tangent line there), the angle the line makes with the horizontal changes by 180°, thus accounting for all slopes of the rotating line. So because every line has a slope (which can be infinity), a line intersecting the oval in some point P also intersects it in some second point Q. The

6.5. Subgroups and Cosets

line can't intersect the oval in three points since the line must intersect the branch, so any line intersects the oval in either zero or two points.

The above observations allow us to see this fundamental group-theoretic fact about the branch and the oval.

> The branch (always including the point at infinity) is a subgroup of the entire real projective elliptic curve.

This is because the points of the branch are closed under addition and subtraction. That's true because the sum of two points in the branch must remain in the branch. If not, that sum would be in the oval, an impossibility because the oval either contains zero or two points, never just one point. Ditto for the difference.

Notation 6.5.1. When a group and subgroup are clear from context, we let G denote the group and H the subgroup. ◇

Definition 6.5.2. A *coset* of a subgroup H of a group G is a translate by some $g_0 \in G$ of that subgroup — that is, a set $H + g_0 = \{h + g_0 : h \in H\}$. We typically assume $g_0 \notin H$. ◇

Comment 6.5.3. It is easy to show that a coset of H is either H itself or is disjoint from it. ◇

Example 6.5.4. $\mathbb{R}^2$ is a group G under vector addition, and any line H through the origin is a subgroup of G. Any (parallel) translate of H by a vector g_0 is a coset of H. Another line through the origin $H' \neq H$ intersects each coset in one point, and we may let that point in the subgroup H' serve as a representative of the coset. The set H' of all these representatives can be thought of as the quotient group G/H of G by H. ◇

Example 6.5.5. In the above example, G can be any $\mathbb{R}^n$ and H, any subspace of G. H' can be any vector space complement of H — for example, the orthogonal complement of H. If H has dimension d, then H' is G/H and has dimension $n - d$. In the expected sense, G is isomorphic to $H \times G/H$ — that is, each element of G can be identified with a pair of elements from the subgroups H and H'. ◇

Example 6.5.6. In Example 6.5.5, the field $\mathbb{R}$ can be replaced by $\mathbb{C}$ or in fact by any field. ◇

Example 6.5.7. Let $G = \mathbb{R}$ and $H = \mathbb{Z}$. Then $\mathbb{R}/\mathbb{Z}$ is isomorphic to a circle group $\bigcirc$. The points of the circle parameterize the cosets of $\mathbb{Z}$ in $\mathbb{R}$, as do the points of the interval $[0, 1) \subset \mathbb{R}$. $\mathbb{R}$ is 1-dimensional. An analogous 2-dimensional example would be $G = \mathbb{R}^2$ and H equal to the plane lattice $\mathbb{Z}^2$. Then $\mathbb{R}^2/\mathbb{Z}^2$ is isomorphic to the torus group $\bigcirc \times \bigcirc$. Its points (or equally well, those of $[0, 1) \times [0, 1) \subset \mathbb{R}^2$) parameterize the points of this quotient group. ◊

Exercise 6.5.8. Establish that any nonvertical line intersects the branch of the cubic $y^2 = x^3 - x$.

Exercise 6.5.9. If $\mathbb{Z}^n$ is a cyclic group of order n and $\mathbb{Z}^m$ is a subgroup of $\mathbb{Z}^n$, when is $(\mathbb{Z}^n)/(\mathbb{Z}^m)$ isomorphic to $\mathbb{Z}^{n-m}$?

Returning to the real curve $y^2 = x^3 - x$ shown in Figure 6.5 on p. 143, we know that the branch is a subgroup H of the entire real elliptic curve G, but what can we say about what's left over, the oval? We've seen before that not only is the oval a loop, the branch is, too — its ends connect at infinity, so the branch forms a topological loop. This suggests that perhaps the oval is a coset of the branch! This is indeed so, and here's why: If g_0 is any point of the oval, we will show that the oval is the set $H + g_0$ — that is, the oval is a group-theoretic translate of the branch by an element of the oval. To see that if g is any point of the oval, then there's an $h \in H$ so that $g = h + g_0$, let $-g$ be the reflection of g about the x-axis. By our previous observations, the line through $-g$ and g_0 intersects H in a point h, so we have $g = h + g_0$, as desired. This is depicted in Figure 6.9.

We've learned that the real projective elliptic curve is a group G, as is its branch H. It's fair to ask what the quotient G/H is. Since H has just the oval as a coset, the quotient consists of two elements, and the only 2-element group is $\mathbb{Z}_2$. The observations above fit in with this, because with $\{0 \leftrightarrow \text{branch}\}$ and $\{1 \leftrightarrow \text{oval}\}$, $0 + 0 = 0$ corresponds to the sum of two branch points being a branch point, while $0+1 = 1+0 = 1$ corresponds to the sum of a branch point and oval point being in the oval. Also, $1 + 1 = 0$ corresponds to the sum of two points in the oval being in the branch.

We are not quite done with our story. Take a look at Figure 6.6 on p. 144 depicting the space curve obtained when our elliptic curve in the complex setting is sliced by the (x_1, y_1, y_2)-subspace of $\mathbb{R}^4$. The part of the space curve that's drawn more lightly is a reflected (and rotated) version of the darker curve we've been discussing. If we look at this lightly

6.5. Subgroups and Cosets

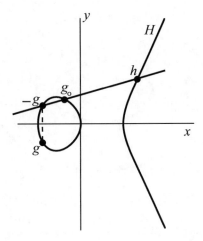

Figure 6.9. The oval is a coset of the real projective curve's branch group. Given points g_0, g in the oval, the line through g_0 and $-g$ intersects the branch in h. This picture can be summed up as $h + g_0 = g$.

drawn curve as simply a plane curve in the usual (x, y)-plane, its equation is obtained from $y^2 = x^3 - x$ by replacing x everywhere with $-x$, giving $y^2 = -x^3 + x$. Just as the branch in $\mathbb{R}^2$ of any curve $y^2 = x^3 + ax + b$ is rightmost and opens to the right, so the branch of any curve $y^2 = -x^3 - ax + b$ is leftmost and opens to the left. Since our definition of addition in elliptic curves uses lines and is purely geometric, the same goes for our reflected version, which tells us that it, too, is an abelian group, say, G'. All our other observations similarly hold, meaning that its branch H' is a subgroup, its oval is a coset of that subgroup, and the quotient G'/H' is the group $\mathbb{Z}_2$.

Figure 6.7 on p. 144 puts these groups, subgroups, and cosets together in a suggestive way as the skeleton of a torus. The two heavily drawn loops together form G, while the other two loops form G'. The outer, heavily drawn loop is the branch subgroup H of G, while the more lightly drawn loop on the left is H' and touches H at infinity, their mutual 0-element. The other two loops are the other two ovals in Figure 6.6 on p. 144 and are cosets of H and H'. To see this more easily, Figure 6.7 is redrawn as Figure 6.10 with additional labeling. Notice that the two heavy loops are disjoint, as are the two lightly drawn loops. In each case, the coset is disjoint from its subgroup.

Chapter 6. Every Real Elliptic Curve Lives in a Donut

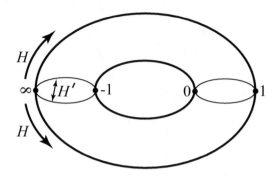

Figure 6.10. The two heavy loops form the group G. The outer heavy loop is the branch subgroup H of G. That branch's 0-element is ∞ in the drawing. The two more lightly drawn loops form the group G'. The left light loop is the branch subgroup H' of G', and that branch's 0-element is again ∞ in the drawing.

Yet another way of relating Figure 6.7 to a skeleton torus is depicted in Figure 6.11. In it, the coordinate pairs appearing in the top and bottom drawings are the endpoints of curves in the top drawing. The curves in the top drawing correspond to the line segments in the bottom drawing in an obvious way. In the bottom drawing, all four corners of the square are labeled $(0,0)$, meaning they're identified to a single point. The two points $(1,0)$ are likewise identified, as are similarly the two points $(0,1)$. The light shading stands for a thin rubber sheet, and the usual way of forming a topological torus from the square by identifying edges automatically makes these identifications. Without any shading, the bottom drawing depicts the real projective curve with its branch subgroup loops, coset loops, and how these touch each other, just as the top picture does. But filling in the square does much more — it corresponds to extending the real projective curve to a complex projective curve. (Although it's customarily called a curve or complex curve, visually it's a real, 2-dimensional surface. This difference is an artifact of nomenclature history.) The locus of $y^2 = x^3 - x$ in $\mathbb{C}^2$, plus the point at infinity, is a closed surface in $\mathbb{P}^2(\mathbb{C})$ — a topological torus. The filled-in square in the bottom drawing has a natural group structure via vector addition in which coordinatewise addition is taken modulo 2, and there is a corresponding group structure

6.5. Subgroups and Cosets

in the surface extending that of the real curve: Addition via the connect-the-dots algorithm is defined geometrically just as in the real case, except now the lines are complex. So, for example, two distinct points in the vector space $\mathbb{C}^2$ determine a unique complex line through them.

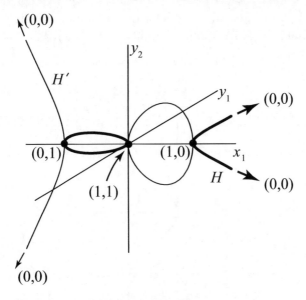

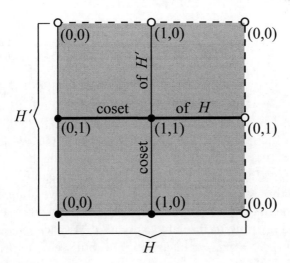

Figure 6.11. Another way our 3D slice relates to a torus skeleton.

6.6 Elliptic Curves with No Oval

Example 6.6.1. The branch and loop in Figure 6.5 are topologically just the loops we tend to draw when making an everyday sketch of a torus. But this chapter's title is "Every Real Elliptic Curve Lives in a Donut," and lots of elliptic curves in the real plane have only a branch and no oval. So, for example, what's the story about an elliptic curve such as $y^2 = x^3 + x$? Its plot in Figure 6.12 reveals no oval.

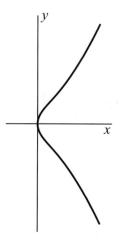

Figure 6.12. This curve $y^2 = x^3 + x = (x + i)x(x - i)$ is analogous to $y^2 = x^3 - x = (x + 1)x(x - 1)$. Every elliptic curve has a loop, but where is it?

If we play the same game as in the last two examples, we get a curve in the 3-dimensional slice $x_2 = 0$ consisting of the branch shown in Figure 6.12, plus its reflection (rotated by 90°) analogous to those examples. That is, the space curve we get is the union of two branches touching at $(0, 0, 0)$, one in the (x_1, y_1)-plane opening to the right, the other in the (x_1, y_2)-plane opening to the left. Now all of 4-space is covered by the parallel 3-slices $x_2 = r$ as r runs through $\mathbb{R}$, but seeing a torus by visualizing the union of the curves in these 3-slices doesn't sound very doable. It would be nice to have a single 3-slice showing loops the way Figure 6.6 does.

6.6. Elliptic Curves with No Oval

How can we get such a slice? A clue is revealed if we work by analogy. Factor the right-hand side of $y^2 = x^3 - x$ into $x^3 - x = (x+1)x(x-1)$. These three factors are 0 when $x = 0$ and ± 1 — the three points where all loops cross the x_1-axis. The loops/branches themselves are formed using x_1 as a parameter. Let's do an analogous thing for $y^2 = x^3 + x$: The right-hand side factors into $(x+i)x(x-i)$, and this is 0 when $x = 0$ and $\pm i$. The loops now cross the x_2-axis (rather than the x_1-axis) at three points, and the loops/branches are formed using x_2 as a parameter. As one example, at $x_2 = \frac{i}{2}$, y^2 is $x^3 + x$ evaluated at $\frac{i}{2}$, which is $\frac{3i}{8}$, and y itself works out to be

$$\pm \frac{\sqrt{3}(1+i)}{4}.$$

These y-values lie in the plane $y_1 = y_2$ in (x_2, y_1, y_2)-space. Similarly, at $x_2 = -\frac{i}{2}$, y^2 is $-\frac{3i}{8}$, and the y-values are

$$\pm \frac{\sqrt{3}(1-i)}{4}.$$

These lie in the plane $y_1 = -y_2$ in (x_2, y_1, y_2)-space. These two planes $y_1 = y_2$ and $y_1 = -y_2$ are perpendicular to each other, in the same sense that $y_1 = 0$ and $y_2 = 0$ are perpendicular in the previous two examples. It turns out that we get a picture looking much like Figure 6.6, just oriented differently in 4-space. ◇

In the previous section we obtained various group-theoretic results for elliptic curves having an oval. Using the ideas just above, we can replicate the geometric arguments of the previous section to get analogous group-theoretic results when there's no oval in (x_1, y_1)-space by using (x_2, y_1, y_2)-space as a 3D slice instead of (x_1, y_1, y_2)-space.

7

The Genus

In this chapter we outline some well-known as well as some not-so-well-known facts about genus. We begin by defining what a 2D closed, orientable topological manifold is. We assume the reader knows the meaning of "compact" and "connected."

7.1 A Few Preliminaries

Definition 7.1.1. (a) A *2D topological manifold* is a surface S such that about every point $P \in S$, there's some S-open neighborhood topologically the same as an open disk in $\mathbb{R}^2$.

(b) A 2D topological manifold S is *closed* provided that its complement in its surrounding space is open. A point P in or not in S is a *boundary* point of S provided P has an open neighborhood homeomorphic to some $x^2 + y^2 < \epsilon$, where $\epsilon > 0$ and $x \geq 0$. In another meaning, S is *closed* if S has no boundary points.

(c) A 2D topological manifold S is *orientable* provided that for any tiny circle supplied with an arrow defining an orientation, whatever path it's pushed along on S, the circle (which we continually flex so it stays in S) returns to its starting position having the same orientation. ◇

Notation 7.1.2. Henceforth, S denotes a compact, connected, orientable 2D topological manifold that is closed in both the above senses. ◇

156   Chapter 7. The Genus

The following is a classification theorem that's fundamental to coffee cup and donut topology. The term "handle" arises from the everyday concept of the handle on a coffee cup. For example, Figure 7.2 contains three handles.

Theorem 7.1.3.

> Any surface S is topologically equivalent to a sphere with some finite number g of handles.

For a proof, see [Massey, Chapter 1, Theorem 5.1].

Definition 7.1.4. The integer g in the above theorem is called the *genus* of the surface.   ◊

7.2 Examples

A Möbius strip is not closed in the second sense in Definition 7.1.1, nor is it orientable. On the other hand, the surfaces in Figures 7.1 and 7.2 are both closed and orientable. They have respective genera 2 and 3. Topologically, we can think of a torus as being a sphere with one handle or a surface with one hole, while Figure 7.1 is a surface with two holes and is topologically the same as a sphere with two handles. Figure 7.2, with its three handles, has genus 3.

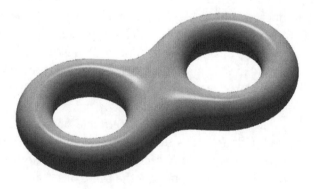

Figure 7.1. A double torus, or a surface of genus 2.

7.2. Examples

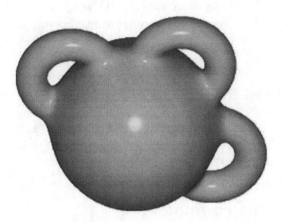

Figure 7.2. A surface of genus 3 as a sphere with three handles.

Comment 7.2.1. It's common to think that depictions of genus as in Figures 7.1 and 7.2 live in ordinary 3-space, but the genus is a topological invariant, so a surface can be stretched, twisted (including into higher dimensions), and even broken apart provided the parts are stitched together so that each "before point" matches up with each "after point." The number of holes never changes under even extreme topological changes. ◊

Comment 7.2.2. As for a polynomial $p(x, y)$ with real and/or complex coefficients, if its coefficients are chosen randomly, then with probability 1, the curve defined by the polynomial has the nice properties that important theorems require. Irreducibility is one and smoothness is another. The argument justifying this is very similar to that in Comment 3.6.2 on p. 69. ◊

From the above comment, we see that with probability 1, any polynomial $p(x, y)$ of degree n with randomly chosen real and/or complex coefficients defines in $\mathbb{P}^2(\mathbb{C})$ a smooth, orientable surface with a finite number g of handles. We'll soon see how g depends on n.

Exercise 7.2.3. A solid wooden ball is centered at the origin of (x, y, z)-space, and holes are drilled all the way through it along each of the three axes. What's the genus of the resulting surface?

Exercise 7.2.4. Redo the above exercise if the solid ball is replaced by a thick wooden shell with outside diameter four inches and inside diameter two inches.

Exercise 7.2.5. In Figure 7.1, cut the right handle to make a torus with two "hoses" coming out of it. Bend one hose downward, guiding it leftward into a concave-up smile, and then feed its end through the remaining hole. Bend the other hose upward and pull its end to the left and reattach it to the end of the first hose. Does "genus" still make sense for this surface? If so, what is it?

7.3 The Genus Formula

We've looked at two examples of elliptic curves, each forming a topological torus when plotted in the complex setting. What about all the many other elliptic curves? It turns out that there is an astonishingly powerful yet simple formula that not only addresses them all but goes much further. Here is what it says:

The Genus Formula:

> If an irreducible polynomial $p(x, y)$ of degree n has real and/or complex coefficients and defines a nonsingular curve $C \subset \mathbb{P}^2(\mathbb{C})$, then as a topological surface, C has genus
> $$g = \frac{(n-1)(n-2)}{2}. \qquad (7.1)$$

For a proof, see Ch. II, Section 10 in [Kendig 3]. This formula is truly sweeping. It says, for example, that any of the six cubic curves $C(p)$ shown in Figure 3.3 on p. 58, when extended via p to a surface in $\mathbb{P}^2(\mathbb{C})$, forms a topological torus. This is because the degree is 3, so the genus is $g = \frac{(3-1)(2-1)}{2} = 1$. It even says that if you randomly pick real or complex coefficients a_i in the general third-degree equation

$$a_0 x^3 + a_1 x^2 y + a_2 x y^2 + a_3 y^3 + a_4 x^2 + a_5 xy + a_6 y^2 + a_7 x + a_8 y + a_9 = 0,$$

then the equation's plot in $\mathbb{P}^2(\mathbb{C})$ is a torus.

The formula goes still further, because it works just as well in extreme cases. If $p(x,y)$ has degree 100, then the number of coefficients of p is

$$1 + 2 + \cdots + 101 = 5{,}151.$$

If these coefficients are chosen randomly, the curve (actually a surface in the complex setting) will with probability 1 be smooth, and the formula says that in $\mathbb{P}^2(\mathbb{C})$, it's a topological sphere with $\frac{(100-1)(100-2)}{2} = 4{,}851$ handles! In an even more extreme case, choosing random coefficients to make a polynomial of degree 1,000 means that the number of handles is 498,501.

Exercise 7.3.1. Randomly pick an integer n, $100 \leq n \leq 200$. What are the chances that some nonsingular algebraic curve in $\mathbb{P}^2(\mathbb{C})$ has genus n?

7.4 The Genus vs. Number Theory

Our basic method in solving a homogeneous number theory problem has been to associate the problem with a nonsingular real algebraic curve defined by a polynomial $p(x,y)$ with rational coefficients. Then, by narrowing our perspective and determining the nature of only the rational points in this curve, we essentially solve the number theory problem. But by broadening our perspective and working in the complex setting, this polynomial $p(x,y)$ also defines a real surface S in $\mathbb{P}^2(\mathbb{C})$ with a genus which we can evaluate through our formula. If we could visualize in four dimensions, that number would just be the number of holes we'd see.

What's so remarkable is that the topology of the surface, which looks at the entire surface in $\mathbb{P}^2(\mathbb{C})$, is intimately connected with what happens at the other extreme — the ultra-thin microcosm of rational points. In the quadratic case the surface's genus is 0 and as we saw in the first two chapters, either there are no solutions or the solutions are parameterized by $\mathbb{Q} \cup \{\infty\}$. If there's one hole ($g = 1$), we've seen that all sorts of complicated things can happen. But if there are two or more holes, the number of possibilities suddenly collapses to just a finite number of solutions — that's the essence of the Mordell Conjecture. This famous conjecture remained just that for over six decades, until Gerd Faltings proved it in 1983–1984. This major result is now often called Faltings's Theorem. Here is one form of it.

Theorem 7.4.1 (Faltings's Theorem).

> Any nonsingular rational algebraic curve of genus $g \geq 2$ has only finitely many rational points.

Comment 7.4.2. A 2013 result of Bhargava ([Bhargava 5]) adds to what we know about nonsingular rational algebraic curves of genus $g \geq 2$ and having the form $y^2 = f(x)$, where $f(x)$ is a polynomial of even degree. (These curves are "hyperelliptic.") Here is the main result:

Theorem 7.4.3.

> The density of those genus-g hyperelliptic curves over $\mathbb{Q}$ having no rational points approaches 100% as $g \to \infty$.

This remarkable result tells us that for curves of high genus, the curve very likely avoids passing through even one rational point. ◊

Although a high-degree number theory problem leads to a high-degree curve that can seemingly perform all sorts of acrobatics in the real plane, these curves manage to miss nearly all the rational points in $\mathbb{R}^2$. This is one of the big mysteries of number theory. A higher degree means the curve does more turning and wiggling, resulting in a higher genus. But since there are so often fewer solutions to number theory problems, it almost tempts one to think that curves twist and turn not to encounter rational points, but rather to *avoid* them!

Let's summarize the connections between genus and how many rational points there are on various nondegenerate curves.

7.4. The Genus vs. Number Theory

Theorem 7.4.4.

> (1) If the genus is 0, then the degree of $p(x,y) \in \mathbb{Q}[x,y]$ is 1 or 2, and the curve is either a line $ax + by = c$ which is nondegenerate in the sense that not both a, b are zero, or it is a nondegenerate conic — that is, an ellipse, parabola or hyperbola. A rational projective line always contains a topological copy of $\mathbb{Q} \cup \{\infty\}$. Any rational projective conic either contains no rational points or a topological copy of $\mathbb{Q} \cup \{\infty\}$. In fact, these rational points are evenly distributed in the sense of being dense within the conic.
>
> (2) If the genus is 1, then the degree of $p(x,y)$ is 3 and the curve is elliptic. Its set of rational points forms a finitely generated abelian group with group structure given by Theorem 4.2.2 on p. 92. The rank of the curve is the rank of the curve's group.
>
> (3) If the genus is greater than 1, then the degree of $p(x,y)$ is ≥ 4. Theorem 7.4.1 (Faltings's Theorem) then guarantees that there can never be more than a finite number of rational points on the curve. Bhargava's result (Theorem 7.4.3) further tells us that for hyperelliptic curves, as the genus g increases without bound, the chance of that finite number actually being zero increases toward 100%.

Comment 7.4.5. We know Mazur's Theorem in Section 4.4 tells us that there are only 15 possible torsion subgroups of an elliptic curve's group. Generally, the more elements in such a subgroup, the rarer they are. When the genus is > 1, there are only finitely many rational points, a 1922 conjecture originally due to Mordell. André Weil's Ph.D. advisor Jacques Hadamard suggested to his student that he try proving it, but young Weil soon saw the writing on the wall and decided against entering a battle he was likely to lose. Indeed, it took over 60 years before Faltings established it, earning him the 1986 Fields Medal. It is now often called Faltings's Theorem. ◇

7.5 The Curious Story of Plane vs. Fancy Curves

Our favorite elliptic curve $y^2 = (x+1)x(x-1)$ passes through $x = -1$, $x = 0$, and $x = +1$. Its affine and projective plots appear in Figure 6.5 on p. 143, with its skeleton and filled-in pictures appearing in Figures 6.7 and 6.8. It turns out that we can generalize from this genus-1 curve to curves of higher genus by just adding more pairs of pass-through points. For example, the curve $y^2 = (x+2)(x+1)x(x-1)(x-2)$ additionally goes through $x = -2$ and $x = +2$. In its affine plot, depicted in Figure 7.3, we see an additional loop, making for two loops and a branch instead of one loop and a branch.

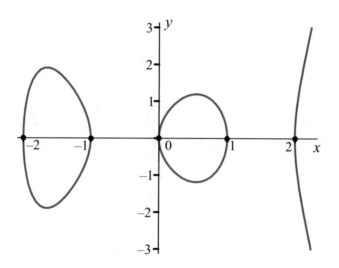

Figure 7.3. Plot of $y^2 = (x+2)(x+1)x(x-1)(x-2)$.

The two loops and branch can be topologically massaged into the skeleton of a double torus, and "longitudes" and "meridians" can be added to form the double torus surface appearing in Figure 7.5.

How does the polynomial $y^2 = (x+2)(x+1)x(x-1)(x-2)$ defining our surface of genus 2 fit in with the genus formula $g = \frac{(n-1)(n-2)}{2}$ in (7.1) on p. 158? The polynomial's degree is 5, but plugging this into the formula gives $g = \frac{(5-1)(5-2)}{2} = 6$ — *way* off! What's going on?

7.5. The Curious Story of Plane vs. Fancy Curves

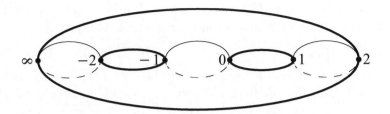

Figure 7.4. Here, the three curves in Figure 7.3 are topologically redrawn as three heavily drawn loops forming part of a double-torus skeleton. The branch in Figure 7.3 passes through 2 and $\{\infty\}$ and is the outermost curve in this skeleton.

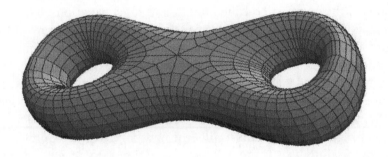

Figure 7.5. Here, the skeleton in Figure 7.4 is filled in with "longitudes" and "meridians" to form a double torus.

Our genus formula works for any nonsingular plane algebraic curve. Certainly the curve is planar and algebraic, so it must not be nonsingular. It's not too hard to show that each of its points in the affine plane is nonsingular, so it must be the point at infinity that's singular. To actually see what the curve around that singular point looks like, homogenize the curve's equation to get

$$y^2 z^3 = x(x^2 - z^2)(x^2 - 4z^2).$$

Dehomogenizing this at $y = 1$ gives

$$z^3 = x(x^2 - z^2)(x^2 - 4z^2).$$

To remove the bending (which is like tremendously magnifying about the singularity), drop all but the lowest-degree terms. This gives
$$z^3 = 0.$$
Since our genus-2 curve $y^2 = (x + 2)(x + 1)x(x - 1)(x - 2)$ isn't nonsingular, does that mean Faltings's Theorem says nothing about it? In fact, as we plug in natural numbers for n in the genus formula $g = \frac{(n-1)(n-2)}{2}$, we get the familiar sequence 0, 1, 3, 6, 10, 15, 21, ..., which has all sorts of gaps! Yet just a minute ago we said, "Our genus formula works for any nonsingular plane algebraic curve." And there, the word *nonsingular* was important. Now it is time to realize that the word *plane* is important, too. Faltings's Theorem doesn't say plane. Here's the story. There are two really big facts about algebraic curves. First, if the curve has singularities, it is possible to "desingularize" the curve using special transformations of the curve — so-called birational transformations. These typically map the curve into higher-dimensional spaces. (For more on these, see, for example, Chapter 5 of [Kendig 2].) And second, birational transformations never change the genus of the curve. So although the curve $y^2 = (x + 2)(x + 1)x(x - 1)(x - 2)$ is a double torus with a singularity, after an appropriate sequence of birational transformations to desingularize, the new nonsingular curve will then be in a space of higher dimension. Faltings's Theorem tells us that the new curve will have only finitely many rational points. There will be more rational coordinates since the space has a higher dimension, so the associated rational polynomial correspondingly has more rational variables than just x and y. Previously, the number of coordinates was two since the rational algebraic curve was assumed to lie in a plane, and the associated number theory problem had three integer variables which we always called a, b, and c. Now there may be more integer variables.

8

In Conclusion . . .

We now summarize and add some perspective to the high points of our journey through homogeneous Diophantine equations of degree 1, 2, and 3. One big takeaway is that as the degree of the Diophantine equation increases, so does the complexity of the answer. For example, in the degree-one situation of $Aa + Bb = Cc$, the general solution is simple and straightforward. In degree 2, if we avoid degenerate cases, then there are only two possibilities — there can be a countably infinite set of solutions or no solutions at all. Today, the world of degree 3 still keeps a sizable group of mathematicians busy trying to penetrate its enduring mysteries. In summarizing, we begin with the case of degree 1 and then move to degrees 2 and 3.

8.1 Degree 1

The degree-one problem $Aa + Bb = Cc$ has a solution

$$\mathbf{a} = \mathbf{Bp},$$
$$\mathbf{b} = \mathbf{Cq} - \mathbf{Ap},$$
$$\mathbf{c} = \mathbf{Bq}$$

when $B \neq 0$, as we saw on p. 35. A, B, C, p, q are all integers with A, B, C given by the problem, and the only assumption is that $Ax + By = C$ ($x = \frac{a}{c}, y = \frac{b}{c}$) actually defines a line. There's a companion solution when $A \neq 0$. (See Exercise 1.13.3 on p. 36.) We needn't worry about "degenerate cases" or the possibility of no solutions encountered with quadratics. It is easy to see from the displayed solution that any slope $\frac{p}{q} \in \mathbb{Q} \cup \{\infty\}$ yields a solution.

8.2 Degree 2

With quadratics, we meet restrictions not arising in the degree-one case. In a degree-two problem, after dividing through by, say, c^2 and writing $x = \frac{a}{c}, y = \frac{b}{c}$, we get a quadratic equation which in its most general form is
$$Ax^2 + Bxy + Cy^2 + Dx + Ey + F = 0.$$
This can define a variety of objects in the (x, y)-plane, including the "nondegenerate conics" — ellipses, parabolas, and hyperbolas. The others consist of the "degenerate conics" which in the real plane look like two lines, one line, a point, or the empty set. In what we've said about finding rational points in the case of degree 2, we have put the spotlight on the nondegenerates and ignored the rest. Why? The reason is that in each degenerate case we can tell, with a little thought, what all the possible rational points are. Here are a few examples of determining the set of rational points in degenerate cases.

Example 8.2.1. $x^2 = 2$. The solutions are $x = \sqrt{2}$ and $x = -\sqrt{2}$. Each solution defines a vertical line in the (x, y)-plane, and each has the irrational number $\sqrt{2}$ or $-\sqrt{2}$ as first coordinate, so both coordinates cannot be rational. Therefore in $\mathbb{Q}^2$, the locus of $x^2 = 2$ is empty, meaning there are no integer solutions to the original problem $a^2 = 2c^2$. ◊

Example 8.2.2. $x^2 = 4$. The solutions are $x = \pm 2$ defining two vertical lines having x-coordinates ± 2. In each case, a rational y-coordinate means the corresponding point is rational and therefore leads to a solution of the problem $a^2 = 4c^2$. Of course, b doesn't appear in the original problem, so every triple $(a, b, c) = (2, b, 1)$, $b \in \mathbb{Z}$, is a solution. ◊

Example 8.2.3. $3x^2 - y^2 = 0$ can be written as $(\sqrt{3}x + y)(\sqrt{3}x - y) = 0$, which defines two lines through the origin of slopes $\pm\sqrt{3}$. Any nonzero rational choice for an x-coordinate means the y-coordinate on either line is irrational since a nonzero rational times the irrational $\sqrt{3}$ is irrational.

8.2. Degree 2

Therefore the solutions to the number theory problem $3a^2 - b^2 = 0$ are the triples $(a, b, c) = (0, 0, c)$ with c any integer. ◊

Example 8.2.4. $(x-3)^2 + (y-4)^2 = 0$ has the unique solution $x = 3$, $y = 4$. Therefore for any integer c, there's just one corresponding solution, $a = 3c$, $b = 4c$. ◊

Exercise 8.2.5. Can a degenerate degree-two object in the real plane ever consist of exactly two distinct points? Why or why not?

We've suggested why less attention is given to degenerate degree-two cases. Now in Diophantine problems of degree 2 whose associated curves are nondegenerate — ellipses, parabolas, and hyperbolas — we've seen that there are exactly two possibilities. To the original number theory problem, either there are no solutions at all or there is a countably infinite set of them, and the set of corresponding points (x, y) can always be parameterized by $\mathbb{Q} \cup \{\infty\}$. Geometrically, within the ellipse, parabola, or hyperbola, the set of rational points is either empty or it's a countable, everywhere dense set within the conic.

So which is which? The answer is straightforward: Find a rational point P on the curve. Then any line of rational slope passing through P intersects the curve in another rational point, and all rational points in the curve are obtained this way. In fact, different slopes produce different rational points and in this way, $\mathbb{Q} \cup \{\infty\}$ parameterizes all the infinitely many rational points on the curve. Often, finding such a rational point P is easy, but when it's not, the simple search program in Appendix C can quickly search through many possibilities. Setting $k = 100$, for example, tests a few million different integer triples. If that doesn't work, set k larger, perhaps to a thousand. This searches through billions of different triples, and if the program still prints out no solutions, the suspicion may well be that there are no solutions. If that's the case, Legendre's Criterion on p. 11 removes all doubt for any curve of the form $Ax^2 + By^2 = C$. [Aitken] goes into the detail of deriving Legendre's Criterion. A more refined and powerful search criterion can be found in [Cremona].

When the conic has rational points, we can think of $\mathbb{Q} \cup \{\infty\}$ as a very narrow decal that gets pasted onto the conic. This can be done in infinitely many different ways. There are a countable number of choices for a rational point P in the conic, and each one can determine where the decal's point 0 goes. We can even say something about what the decal looks like. Pass a line through the center of a circle. That line has a slope and intersects the circle in two diametrically opposite points. By identifying each pair of opposite points to one point, we get a new topological circle, and the lines' slopes paint a copy of $\mathbb{R} \cup \{\infty\}$ — and therefore also of $\mathbb{Q} \cup \{\infty\}$ — onto this new circle. Now each conic in the projective plane is itself a circle, so pasting the circular decal onto the conic amounts to pasting one circle onto another!

8.3 Degree 3

The basic thrust of this book is that a homogeneous Diophantine equation gets converted to a polynomial equation $p(x, y) = 0$ with, say, $x = \frac{a}{c}$ and $y = \frac{b}{c}$. We then try to characterize the rational points of the plane curve defined by $p(x, y) = 0$. The number and types of possible plane curves already hint at how much more intricate degree-three problems are than degree-one or degree-two problems. When the degree is 1, we're looking at just a line, and for degree 2, simply an ellipse, parabola, or hyperbola. But degree 3? This is the case that has kept more mathematicians busy for more years than any other case, and the possible solutions to a homogeneous Diophantine equation of degree 3 in unknowns a, b, c splinter into a multitude of shards. Geometrically, we see just a few of these pieces in Figure 3.3 on p. 58. On p. 57, the general third-degree polynomial equation in (3.1) has ten coefficients and fills up an entire printed line. Figure 3.3 hardly does justice to the range of curves, and mathematicians initially working on degree 3 must have asked themselves, "Where to start?"

Game-changing advances came through massaging these curves using "birational transformations" (see [Kendig 2]) to cast the curves in a more standard form with their polynomials having far fewer coefficients — the Weierstrass long form with only five coefficients:

$$y^2 + a_1 xy + a_3 y = x^3 + a_2 x^2 + a_4 x + a_6,$$

and the Weierstrass short form with just two coefficients:

$$y^2 = x^3 + ax + b.$$

These forms made viewing the possible curves incredibly easier to understand. As an example, compare the small scattershot sampling of curves in Figure 3.3 on p. 58 with the well-behaved suite of elliptic curves in Figure 8.1.

8.3. Degree 3

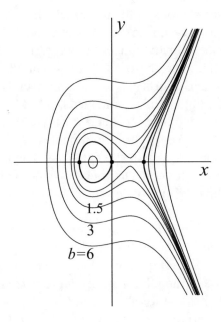

Figure 8.1. Nine examples of elliptic curves in Weierstrass short form.

These nine curves do a credible job of depicting the shape of typical short-form elliptic curves. Any one of these elliptic curves can be thought of as a representative of infinitely many elliptic curves birationally equivalent to it — and not only a representative, but the "very nicest" representative. It's not only pleasant to look at, but y occurs only as a square in the short form $y^2 = x^3 + ax + b$, meaning that its curve is symmetric about the x-axis. This is important, since in defining the sum of points P_1 and P_2 on a short-form elliptic curve, the third point on the curve lying on the line through P_1 and P_2 is reflected about the x-axis.

Long-form curves are also important. The coefficients are never larger than in the short form and can be dramatically smaller. Figure 4.9 on p. 99 provides a look at some long-form curves. A nonzero xy-term can twist or rotate the curve, and we see this in (e) of the figure. In the other parts (f), (g), and (h), the x^2-term appears with a nonzero coefficient.

We mentioned just above that the full solution to a degree-three problem tends to splinter into smaller parts. Although the set of all rational points of any elliptic curve obligingly forms an abelian group, it is nonetheless necessary to look at the pieces which are comprised of a torsion group and/or a finite number of copies of $\mathbb{Z}$, all joined together by direct sum. Fortunately, for an impressively large number of elliptic curves, we now have specific answers to what the set of all rational points is, thanks to the Herculean efforts of many mathematicians. As of this writing, the number of elliptic curves cataloged is 3,064,705, and anyone can freely access this huge database by typing www.LMFDB.org/ into a search engine such as Google. We noted in Comment 5.5.1 just why such a specific number arises as the size of the database: The number of rational elliptic curves of conductor equal to or less than 500,000 comes to exactly 3,064,705. A thumbnail of instructions on how to use the database is given on pp. 102–103.

Earlier, we pointed out that as the rank or torsion group size increases, the associated curves become more rare; you can easily use LMFDB to find specific results. As one example, out of all those curves (3,064,705 as of this writing), only six of them have a torsion group of size 16. That group is $\mathbb{Z}_2 \oplus \mathbb{Z}_8$. As another example, out of those 3,064,705 curves, there is only *one* of rank 4 — the curve $y^2 + xy = x^3 - x^2 - 79x + 289$. Ask for both rank and torsion, and examples becomes even rarer: Not a single curve in the database has rank 3 and torsion $\mathbb{Z}_5$. With LMFDB, you can find a multitude of similar examples on your own.

By the way, although LMFDB sketches every one of the 3,064,705 curves, not a single one of the sketches in Figure 3.3 on p. 58 will be found among them. All 3,064,705 curves are for elliptic curves given in long or short form, while all the curves in Figure 3.3 were plotted by randomly choosing the ten coefficients in the general cubic in variables x and y.

We have now concluded this book's brief journey into the juncture of number theory with algebraic curves. On the theoretical side, important modern advances are continually being established — think of how theory helped crack Fermat's Last Theorem, which withstood attempts at proof for some 350 years. Theories tend to build upon each other, making our exploratory machine ever more powerful. On the calculation and data-gathering side of the ledger, today's rapid advances toward quantum computing promise incredibly vast amounts of data from which mathematicians can form new conjectures, and these can then be subjected to theory.

I hope this journey has been pleasant and profitable!

Appendix A

What Is a Smooth Complex Curve?

We argued in Section 3.4 that smoothness should be defined in the complex setting to avoid unexpected glitches and exceptions. So exactly what *does* smoothness mean in the complex setting? For a curve defined by $p(x,y) = 0$, the definition still comes down to at least one of p_x, p_y being nonzero. This in turn rests on one of the most fundamental results in analysis, the Implicit Function Theorem. This gives a sufficient condition for the zero set of a function F to locally be the graph of some analytic function. For this book, we take F to be a polynomial $p(x,y)$ of two variables.

Definition A.1. A differentiable complex-valued function is *analytic at a point* if and only if the Taylor series about that point converges to the function throughout some neighborhood of the point. The curve $p(x,y) = 0$ is then called *smooth* at a point if the curve is the graph of some function analytic at that point. If the curve $p(x,y) = 0$ is smooth at each point, we call the curve itself *smooth*. ◊

We now give the theorem leading to the simple test for smoothness in Definition A.3.

Theorem A.2 (The Implicit Function Theorem).

> Let $(x_0, y_0) \in \mathbb{C}^2$, and suppose the polynomial p satisfies $p(x_0, y_0) = 0$ and $p_y(x_0, y_0) \neq 0$. Then within some sufficiently small neighborhood of (x_0, y_0), the solutions of $p(x, y) = 0$ form the graph of a uniquely defined function $y = f(x)$ analytic at x_0.

The proof of the Implicit Function Theorem is a bit long and is typically given for analytic functions of more variables and higher dimensions rather than for only one polynomial of two variables. Three references are [Fischer, Appendix 3, pp. 193–196], [Griffiths, Chapter I, §9], and [Whitney, Appendix II, §3]. This theorem is sometimes more fully referred to as the Holomorphic Implicit Function Theorem.

The conditions of the Implicit Function Theorem are algebraic, but in the real setting we can outline some geometric intuition. First, $p(x_0, y_0) = 0$ and $p_y(x_0, y_0) \neq 0$ can be simplified to $p(0, 0) = 0$ and $p_y(0, 0) \neq 0$ by replacing x by $x + x_0$ and y by $y + y_0$. We can also be more specific about the form of $p(x, y)$. Since $p(0, 0) = 0$, we can write

$$p(x, y) = ax + by + \text{terms of degree} \geq 2.$$

Notice that the partials with respect to x or y of every term of degree ≥ 2 are zero at $x = y = 0$. Therefore $p_y(0, 0)$ can be nonzero only if $b \neq 0$, meaning we can write $ax + by = 0$ as $y = \frac{-a}{b}x$. Note that within a sufficiently small neighborhood U of $(0, 0) \in \mathbb{R}^2$, the contribution of the terms of degree ≥ 2 is negligible compared to that of $ax + by$ (the square of a thousandth is a millionth, the cube of a thousandth is a billionth, and so on), so U can be chosen so small that the graph of $y = \frac{-a}{b}x$ is visually indistinguishable from that of $p(x, y) = 0$ there. That is, the zero set of $p(x, y)$ within U looks just like the line $y = \frac{-a}{b}x$, which is a function. If $p_x(0, 0) \neq 0$, the above argument is the same, with x and y interchanged.

Intuitively, a real curve like a line or a nondegenerate conic is smooth everywhere. On the other hand, the crosspoint of a real alpha curve means that real curve is not smooth. In the complex setting, the part of a complex alpha curve in some $\mathbb{C}^2$-open neighborhood of the crosspoint is topologically two open disks touching at just that one point, shown in the top drawing of Figure A.1. However, the origin of the cusp curve $y^2 = x^3$ is

Appendix A. What Is a Smooth Complex Curve?

different, because although the part of the curve in a neighborhood of the origin is *topologically* a disk in $\mathbb{C}^2$, it is contorted so much that in no rectangular coordinate system is the part within a neighborhood of the origin the graph of an analytic function. This is shown in the bottom drawing of Figure A.1.

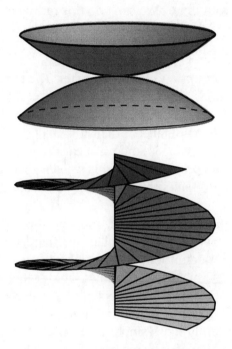

Figure A.1. The top figure represents two topological disks touching at one point, as happens with $x^2 = y^2$ about the origin of $\mathbb{C}^2$, while the bottom figure represents one topological disk corresponding to the part about the origin of the cusp curve $y^2 = x^3$. The figure's topmost edge is identified with its bottommost edge, and in the four real dimensions of $\mathbb{C}^2$, they are glued together without self-intersection.

Although the Implicit Function Theorem gives a sufficient condition for a point of an algebraic curve $p(x, y) = 0$ to be smooth, it's not a necessary one. For example, the parabola defined by $p(x, y) = x - y^2$ is smooth everywhere, but $p_y(0, 0)$ isn't nonzero, so the theorem yields no information there. However, we could equally well state the theorem singling out x instead of y. In this form the theorem would tell us that if $p(0, 0) = 0$

and $p_x(0,0) \neq 0$, then the part of the parabola around the origin is the graph of a uniquely defined function $x = g(y)$ analytic at $y = 0$. If at any point P of any algebraic curve $p(x, y) = 0$ at least one of p_x and p_y is nonzero, we could apply one or the other form of the theorem to conclude that the curve is locally the graph of some function complex-analytic at P and therefore smooth there. We can't do this for the cusp curve $y^2 = x^3$ since for $p = y^2 - x^3$, both p_x and p_y are zero at the origin.

This leads to an alternative version of Definition A.1.

Definition A.3. Let $p(x, y)$ be irreducible. The curve $p(x, y) = 0$ in $\mathbb{C}^2$ is *smooth* at (x_0, y_0) if and only if $p(x_0, y_0) = 0$ and at least one of $p_x(x_0, y_0)$ and $p_y(x_0, y_0)$ is defined and nonzero. If the affine curve is smooth at each of its points in $\mathbb{C}^2$, then it is *smooth* in $\mathbb{C}^2$. If a curve in the complex projective plane is smooth at each of its points in each affine part, then we say the projective curve is *smooth*. The term *nonsingular* may be substituted for "smooth" throughout this definition. ◇

Although this definition is correct and any elliptic curve satisfies it at each of its points, there are many real curves that look perfectly smooth but, according to our definition, are not. The reason is that by restricting our vision to the real plane, we can miss seeing what it is that makes a complex curve nonsmooth. Take a look at the "valentine curve" defined by $x^4 + y^4 = y(x^2 + y^2)$ and plotted in Figure A.2.

This curve certainly *looks* smooth, but according to our definition, $(0, 0)$ is a nonsmooth point. The reason is that, in addition to the curve we see, there's an extra isolated singular point at the origin. It doesn't make the real curve look any different, but the complex curve looks wildly different at the origin. The complex tangent line at a nonsingular point is some copy of $\mathbb{C}$, so it looks like a real plane. But at the valentine curve's singular point, instead of one real plane in $\mathbb{C}^2$, there are three different planes going through the origin there.

Since smoothness in the complex setting is necessary for many of our most important number-theoretic methods and results to hold, it would be good to see what Definition A.3 says about the geometry about that point. Because most of us can't see well enough in four dimensions to make reliable statements about what the curve looks like around a singularity, we need a little algebraic help. Fortunately, there exists exactly the kind of help we need. In $\mathbb{C}^2$ there is an easy way to see enough

Appendix A. What Is a Smooth Complex Curve?

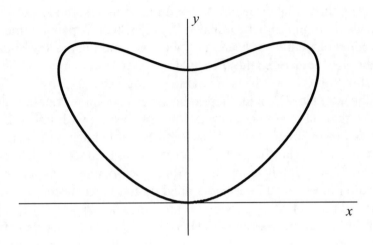

Figure A.2. This "valentine curve" $x^4+y^4 = y(x^2+y^2)$ may look smooth, but it isn't — the origin is a singular point.

of what's happening at a singular point to make accurate statements about that singularity. Here's the basic idea: Replace the actual part of the curve around a point by the tangents to the curve there. For example, the alpha curve $y^2 = x^3 + x^2$ has a singularity at the origin, and greatly magnified about the origin, the real curve looks like the two crossing lines $y = x$ and $y = -x$. The singularity where these lines cross has exactly the same nature as the self-intersection point of the alpha curve. In fact, if we magnify enough around the alpha curve's singularity, we wouldn't be able to visually tell the difference between the curve and the two crossing lines.

There is a very simple way to get these two lines directly from the curve's equation:

> Drop all higher-degree terms, leaving only the terms of lowest degree.

For example, in $y^2 = x^3 + x^2$, drop the x^3 term, leaving $y^2 = x^2$, whose solutions are precisely those two crossing lines $y = x$ and $y = -x$. The idea behind this is that near the origin where x- and y-values are small, higher powers of a small number get even smaller, and sufficiently near the origin, their influence in the plotting process becomes negligible, so we can safely drop those higher-degree terms. In keeping just the lowest-degree terms, we assume the polynomial $p(x,y)$ has been expanded, meaning any product of factors has been multiplied out, making the lowest-degree part homogeneous in x and y. In analogy to the Fundamental Theorem of Algebra, this can now be factored into linear pieces $ax + by$, with a and b possibly complex. Each of these represents a complex line through the origin. (As usual, we say "complex line" in keeping with linear algebra usage, but each "line" is actually a complex plane having the usual two real dimensions.) So the part of a curve about a singularity is replaced by the union of finitely many lines. Some factors may be repeated, corresponding to identical lines being piled on top of each other — that is, "lines with multiplicity." Curving is hard to see in 4-space. Dropping the higher-degree terms takes away the part of the polynomial $p(x,y)$ responsible for the curving. Curving itself takes place away from the singularity and only complicates things. We don't need that information, and our method removes it.

It is now time to flesh out these ideas with some examples.

Example A.4. Changing a sign in the alpha curve equation $y^2 = x^2(x+1)$ to $y^2 = x^2(x-1)$ creates an isolated point at $(0,0)$. Certainly $(0,0)$ satisfies this new equation, but for small $x \neq 0$, the $(x-1)$ in $y^2 = x^2(x-1)$ means $y^2 < 0$, so y becomes imaginary as soon as x moves away from zero. This curve is depicted in Figure A.3.

Expanding $y^2 = x^2(x-1)$ to $y^2 = x^3 - x^2$ and dropping the cubic term leaves $y^2 = -x^2$, giving the two lines $y = \pm ix$. Now in $\mathbb{C}^2$ there are visually two different kinds of complex lines $y = mx$. When m is real, we see an ordinary line in the real (x,y)-plane, but when $m \in \mathbb{C} \setminus \mathbb{R}$, we see less — just the origin in the real (x,y)-plane. This last can be justified by looking at m as $m/1$ ($m \in \mathbb{C} \setminus \mathbb{R}$), meaning that the point $(m,1)$ is in the complex line (or plane, if you prefer). An easy argument shows that no nonzero real or complex multiple of $(m,1)$ can have both components real, which is required for a point to be in the real (x,y)-plane. By the way, this shows that of all the complex lines in $\mathbb{C}^2$, if we see a line in the real plane, it must have a real slope. In every other case, we see nothing or just one point in the real plane. ◇

Example A.5. Look again at the valentine curve depicted in Figure A.2. In its equation $x^4 + y^4 = y^3 + x^2 y$, keeping only the lowest-degree terms

Appendix A. What Is a Smooth Complex Curve?

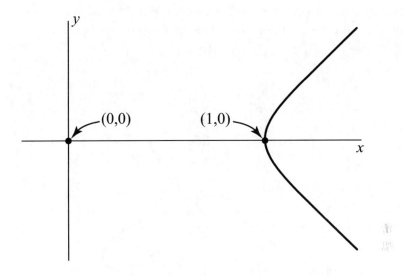

Figure A.3. The origin is a singular point of this curve $y^2 = x^2(x-1)$. The rest of the curve is smooth.

leaves $0 = y^3 + x^2y$. Factoring this gives $y(y + ix)(y - ix) = 0$. This means there are three different tangent lines through the origin, which in turn tells us that the point is singular. Of the three lines, only one appears as a line in the real plane, while each of the other two, having imaginary slopes, appear as the origin there. In $\mathbb{R}^4$, these tangents form a real surface consisting of three planes intersecting in the origin. The higher-degree terms $x^4 + y^4$ make the surface curve away from the planes, but sufficiently close to the origin this deviation is negligible. ◊

Example A.6. The real three-leaved rose commonly found in calculus texts has three distinct real tangent lines through the curve's singular point. In $\mathbb{C}^2$, these three tangents appear as three real planes mutually intersecting in the singularity. As a concrete example, the three-leaved rose in Figure A.4 happens to be defined in rectangular coordinates by $y(y^2 - 3x^2) = (x^2 + y^2)^2$. The left-hand side is homogeneous of degree 3. The right-hand side is homogeneous of order 4, so dropping it leaves $y(y^2 - 3x^2) = 0$. Since $y(y^2 - 3x^2)$ factors into $y(y + \sqrt{3}x)(y - \sqrt{3}x)$, we see that the tangent lines have slopes 0 and $\pm\sqrt{3}$ — that is, lines making 60° to each other. These three lines through the origin are depicted in Figure A.4.   ◊

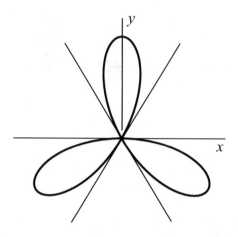

Figure A.4. The part of this rose, when greatly magnified around the origin, looks like the three depicted tangent lines.

By the way, the $(x^2 + y^2)^2$ in the rose's equation forces the tangent lines to bend to keep the whole real figure bounded. That's because as x and y become sufficiently large, $(x^2 + y^2)^2$ grows larger than values of the polynomial of smaller power, so at all points (x, y) sufficiently far away from the origin, the two sides of the rose equation can never equal each other. With no solutions there, the figure must be bounded. Other large even powers do the same thing — powers such as $(x^2 + y^2)^3$, $x^4 + y^4$, and so on. Larger powers will increasingly tend to flatten out the petal tips so they are more squarish. More generally, adding large even powers bounds the plot of any polynomial of lower degree.

Exercise A.7. Find a polynomial defining a curve looking like the teardrop curve in Figure A.5.

Example A.8. The cusp curve depicted on the right in Figure A.6 has a singularity at the origin. The tangent line to the curve there can be regarded as two lines on top of each other — the limit of the two tangent

Appendix A. What Is a Smooth Complex Curve?

Figure A.5. A teardrop curve.

lines at the origin of the alpha curve $y^2 = x^2(x+\epsilon)$ as $\epsilon > 0$ approaches 0, suggested by the drawing on the left. Now $(y^2 - \epsilon x^2) = (y + \sqrt{\epsilon}x)(y - \sqrt{\epsilon}x)$ is the lowest-degree part of this, so the tangents' slopes $m = \pm\sqrt{\epsilon}$ approach 0 as $\epsilon \to 0$. ◇

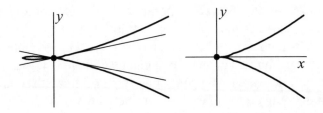

Figure A.6. The cusp on the right is the limit of alpha curves $y^2 = x^2(x + \epsilon)$ as $\epsilon > 0$ tends to 0. This makes the loop size go to zero and the tangents at $(0,0)$ tend to the horizontal.

Example A.9. The curve depicted in Figure A.7 ups the ante of the single cusp curve. This one has two different cusp curves tangent at the origin, so there are now *four* lines atop each other — a line of multiplicity 4. This curve's equation is $(y^2 - x^3)(y^2 - 4x^3) + x^8 + y^8 = 0$, with its lowest-degree part being y^4. Each cusp contributes a double tangent line there, giving the total of four lines. The equation of this curve was constructed by multiplying together the equations of the two cusp curves $(y^2 - x^3) = 0$ and $(y^2 - 4x^3) = 0$. Each cusp curve is unbounded, but we bound this real curve by adding $x^8 + y^8$, which is an even power larger than the largest power of 6 that $(y^2 - x^3)(y^2 - 4x^3)$ has after multiplying out. ◇

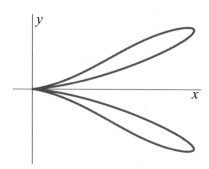

Figure A.7. A curve with two cusp points at the origin.

Example A.10. Figure A.8 is the plot of $(x^2 + y^2)^3 = x^2y^2$, a four-leaved rose. Notice that at the origin, one branch of the curve is concave up, another concave down. This corresponds to the double horizontal tangent $y^2 = 0$ at the origin. Similarly, there's the double vertical tangent $x^2 = 0$ there. The left-hand side of $(x^2 + y^2)^3 = x^2y^2$ is homogeneous of degree 6, while the right-hand side is homogeneous of degree 4, giving the four tangent lines. There are no tangent lines at the origin with nonreal slope, so other than being on top of each other, there are no hidden lines. ◊

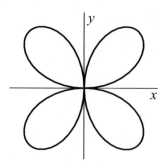

Figure A.8. This rose has double horizontal and vertical tangents at the origin.

Appendix B

Algebraic Curves in the Disk Model

We elaborate here on some of the ideas in Chapter 2, providing more detail and somewhat greater generality. To map an affine curve into the projective disk, we mapped each point (x, y) of the real plane $\mathbb{R}^2$ into that disk via the map

$$(x, y) \longrightarrow \left(\frac{x}{\sqrt{x^2 + y^2 + 1}}, \frac{y}{\sqrt{x^2 + y^2 + 1}} \right).$$

The image of a line L such as $y = 1$ illustrates the shrinking, since the unit vertical distances from L to the x-axis in $\mathbb{R}^2$ decrease under the map. Figure B.1 depicts the image in the open disk of the two horizontal lines $y = \pm y_o$ in $\mathbb{R}^2$. For any fixed $y_o \in \mathbb{R}$, the images of $y = \pm y_o$ lie on an ellipse. Its semimajor axis has length 1, and since $(0, y_o) \in \mathbb{R}^2$ maps to

$$\left(0, \frac{y_o}{\sqrt{y_o^2 + 1}} \right),$$

the semiminor axis of the ellipse has length

$$\frac{y_o}{\sqrt{y_o^2 + 1}}.$$

Appendix B. Algebraic Curves in the Disk Model

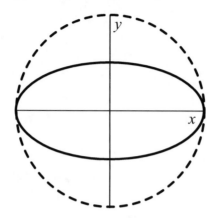

Figure B.1. Two parallel lines $y = \pm y_o$ form an ellipse in the disk model.

It is easily checked that the image of any point $(c, \pm y_o)$ satisfies the equation of this ellipse.

More generally, the images of all mutually parallel lines make up a family of ellipses with the same major axis, but without the two endpoints of this axis. It is natural to add to the picture these two endpoints, but since these two points lie on the boundary of the unit disk, any two of these lines intersect in *two* points rather than one. In Chapter 2 we identified the two opposite, or antipodal, points, and we now motivate this more fully. Pick two lines in the real plane — L (fixed) and another passing through a fixed point Q not on L and intersecting L in the point P, as suggested in Figure B.2.

When we rotate the line containing point Q so that the line approaches being parallel to L, the intersection point P flies off in one direction towards infinity. As we rotate beyond parallelism, the intersection point pops up on the other side of the line, moving in the same direction as before. In this respect the two added points act like one ordinary point, and it is natural to identify them.

Of course, choosing a different fixed line L and point Q leads to a different pair of antipodal points. So we separately identify *each* pair of antipodal boundary points of the open disk. This construction is similar to creating a Möbius strip or torus by gluing together, or identifying,

Appendix B. Algebraic Curves in the Disk Model

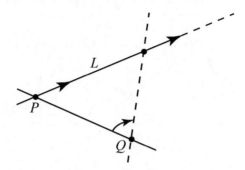

Figure B.2. As the line through Q rotates toward being parallel to L, P travels toward infinity. Rotating beyond that, P appears on the other side of L. When the rotating line is exactly parallel to L, it is natural to create a single point and call it the point at infinity.

appropriate edges of a rectangle. In identifying antipodal boundary points on a disk, we identify a pair of points to a single point. In creating a Möbius strip or torus from a rectangle, there are many such identifications of point-pairs to single points.

Definition B.1. A closed disk with antipodal boundary points identified and supplied with the topology given in Definition B.3 on p. 186 is the *topological disk model of the real projective plane* $\mathbb{P}^2(\mathbb{R})$. ◊

We can apply the same technique to parametrizations of other curves to draw their disk model images.

Example B.2. Figure B.3 depicts the disk model of the alpha curve defined by $y^2 = x^2(x+1)$. It's straightforward to check that the parameterization $t \to (x(t), y(t))$ for this curve is

$$t \to (t^2 - 1, \, t(t^2 - 1)).$$

Here, the antipodal boundary points $(0, 1)$ and $(0, -1)$ of the disk are identified, making the alpha curve image a topological figure 8. ◊

Appendix B. Algebraic Curves in the Disk Model

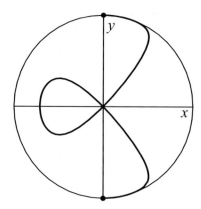

Figure B.3. An alpha curve plotted in the real projective disk.

A Teachable Picture

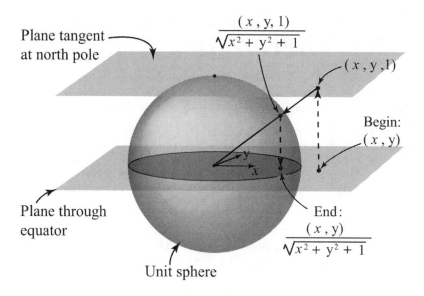

Figure B.4. Follow the arrows to see how a point in the (x, y)-plane maps into the unit disk.

A Teachable Picture

Figure B.4 accomplishes several things. Out of the many ways we could map $\mathbb{R}^2$ to the open unit disk, this picture reveals the thinking behind our particular choice. It also leads to a symmetric definition of $\mathbb{P}^2(\mathbb{R})$ and makes it easy to define a natural topology on $\mathbb{P}^2(\mathbb{R})$. Additionally, the picture provides a way to recenter at infinity, permitting a detailed look at how curves behave there. Finally, it suggests a vector space definition of $\mathbb{P}^2(\mathbb{R})$ which provides an easy way to generalize to $\mathbb{P}^n(\mathbb{R})$ and $\mathbb{P}^n(\mathbb{C})$.

Here's how Figure B.4 reveals the geometry behind our shrinking formula. First, the picture shows the unit sphere centered at the origin of $\mathbb{R}^3$ and the plane $z = 1$ parallel to the (x, y)-plane. Now start at any point (x, y) in $\mathbb{R}^2$, project it vertically to $(x, y, 1)$ in the plane $z = 1$. Notice that $(x, y, 1)$ is a distance $\sqrt{x^2 + y^2 + 1}$ from the origin. Our next move is radially to the sphere, landing at

$$\frac{(x, y, 1)}{\sqrt{x^2 + y^2 + 1}}$$

which, due to the denominator, is now one unit from the origin — that is, on the unit sphere. Dropping down to the (x, y)-plane is the same as dropping the third coordinate, meaning we end up at

$$\frac{(x, y)}{\sqrt{x^2 + y^2 + 1}}.$$

This geometric way of looking at our shrinking map leads to a basic observation. A line in the (x, y)-plane projects up to a line in the plane $z = 1$, which together with $(0, 0, 0) \in \mathbb{R}^3$ defines a plane through $(0, 0, 0)$, and that in turn intersects the upper hemisphere in the top half of a great circle. So the disk image of a typical line is the projection of a great semicircle on the hemisphere.

Vertical projection defines a one-to-one and onto map between the disk model and the *hemisphere model* of the real projective plane: the upper hemisphere with opposite equatorial points identified. That's just a step away from looking at $\mathbb{P}^2(\mathbb{R})$ as the entire sphere in which each pair of antipodal points is identified to a point. This is the *sphere model* of the real projective plane and is beautifully symmetric, but we can go even further and eliminate the arbitrariness in choosing a particular size of sphere. Instead of identifying a point-pair to a point, *identify an entire 1-space to a point*. (A 1-space is any line through the origin.) That is, we regard points of $\mathbb{P}^2(\mathbb{R})$ as 1-subspaces of $\mathbb{R}^3$. In this way, the radius of a sphere centered at the origin becomes irrelevant. $\mathbb{P}^2(\mathbb{R})$ can be looked at as the set of all 1-subspaces of $\mathbb{R}^3$, and we call this the *vector space model* of the real projective plane.

Some Basic Definitions

We have now met four different models of the real projective plane. In the next definition we define a topology on each. For this, recall that a set of basic open sets of $\mathbb{R}^3$ can be taken to be the set of open balls there.

Definition B.3.

- In the vector space model of $\mathbb{P}^2(\mathbb{R})$, the points are the 1-subspaces of $\mathbb{R}^3$. A typical *basic open set* $\mathcal{O}$ for the natural topology on this model consists of all 1-subspaces of $\mathbb{R}^3$ intersecting an open ball of $\mathbb{R}^3$.

- The points of the sphere model of $\mathbb{P}^2(\mathbb{R})$ can be taken as the antipodal point pairs of the sphere $\mathcal{S} : x^2 + y^2 + z^2 = 1$ in $\mathbb{R}^3$. A typical *basic open set* for the natural topology on this model consists of all points of $\mathcal{S}$ intersecting any one basic open set $\mathcal{O}$ in the vector space model.

- The points (x, y, z) of the hemisphere model of $\mathbb{P}^2(\mathbb{R})$ can be taken to be those of the sphere $\mathcal{S}$ for which $z \geq 0$, with antipodal equatorial points ($z = 0$) being identified. A typical *basic open set* for the natural topology on this model is the intersection of the hemisphere with a basic open set of $\mathcal{S}$.

- The points of the disk model of $\mathbb{P}^2(\mathbb{R})$ can be taken to be the projections $(x, y, z) \to (x, y)$ of points in the hemisphere model to points in the disk $x^2 + y^2 \leq 1$. A typical *basic open set* for the natural topology on this model is the projection of a basic open set of the hemisphere. ◇

To take advantage of viewing points of $\mathbb{P}^2(\mathbb{R})$ as 1-subspaces of $\mathbb{R}^3$, we need to determine what an algebraic curve C in $\mathbb{R}^2$ looks like in this model and how we then add the points of infinity to the curve. Let's start with a basic example, a line L in $\mathbb{R}^2$.

Example B.4. Vertically lift a line L in $\mathbb{R}^2$ to a line L' in the plane $z = 1$ in $\mathbb{R}^3$. A point P in L' determines a 1-space of $\mathbb{R}^3$ through P, and therefore a point in the vector space model of $\mathbb{P}^2(\mathbb{R})$. Within the plane $z = 1$,

Some Basic Definitions

the farther away P is from the origin, the smaller the angle between the 1-space through P and the (x, y)-plane in $\mathbb{R}^3$. For points of L', this angle never quite reaches 0, so the set of 1-subspaces forms the plane through L' and the origin of $\mathbb{R}^3$, minus L. Of course "the angle reaching 0" would mean that the 1-subspace is parallel to the plane $z = 1$, which corresponds to P being at infinity. By adding this line L, we get the entire plane through L' and the origin of $\mathbb{R}^3$. Thus, a projective line in $\mathbb{P}^2(\mathbb{R})$ is represented by a 2-space in $\mathbb{R}^3$. This, together with what we've seen above, shows that in $\mathbb{R}^3$, subspaces of dimension 1, 2, and 3, respectively, correspond in $\mathbb{P}^2(\mathbb{R})$ to points, lines, and all of $\mathbb{P}^2(\mathbb{R})$. In fact, the 0-dimensional subspace of $\mathbb{R}^3$ — the origin of $\mathbb{R}^3$ — defines the empty set in $\mathbb{P}^2(\mathbb{R})$, to which we assign dimension -1. To sum up:

> Subspaces of dimension n in $\mathbb{R}^3$ correspond to projective subspaces of $\mathbb{P}^2(\mathbb{R})$ of one lower dimension.

In topological terms, adding L to the union of the 1-subspaces through L' corresponds to taking the topological closure of the union, with L''s point at infinity being the 1-subspace L. This closure is a *homogenization* of L' in $\mathbb{R}^3$ and we denote it by $H(L')$. It is homogeneous in the usual sense — if it contains a nonorigin point Q, then the 1-subspace containing Q is in it, too. ◇

Example B.5. Let C be a curve in the real (x, y)-plane defined by a polynomial of degree 2. Lifting this curve to the plane $z = 1$, passing 1-spaces through each point, and then taking the topological closure homogenizes C. Doing this to the circle C defined as the zero set of $(x - 1)^2 + y^2 - 4$, for example, produces the cone illustrated in Figure B.5. The cone is the zero set of the homogenization of $(x - 1)^2 + y^2 - 4$. The cone's equation is therefore $x^2 - 2xz + y^2 = 3z^2$. ◇

Modeling $\mathbb{P}^2(\mathbb{R})$ either as a sphere with antipodal points identified or as the 1-subspaces of $\mathbb{R}^3$ has another important consequence: The equator no longer plays a special role. This will allow us to recenter anywhere in the projective plane, even at a point at infinity. Recentering at any point P in the projective plane allows us to write the equation of the curve so that P is the new origin, allowing us to see in a precise way the behavior of the curve there.

Appendix B. Algebraic Curves in the Disk Model

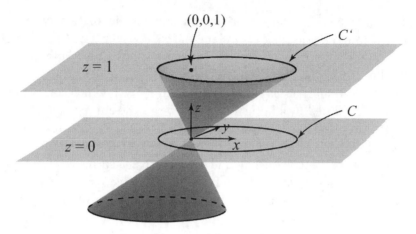

Figure B.5. Homogenizing C by passing 1-subspaces of $\mathbb{R}^3$ through points of C'.

Example B.6. We can illustrate this recentering idea using Figure B.5. The lifted circle C' generates the cone, and we can think of the cone as a representation of the circle in $\mathbb{P}^2(\mathbb{R})$. We could slice the cone with the affine plane or viewing screen $z = 1$ to get the circle again, but there's no special reason to select $z = 1$ as the screen. We could just as well choose any other plane not passing through the origin of $\mathbb{R}^3$. As we choose various viewing screens this way, we encounter a variety of conic sections — circles, ellipses, parabolas, and hyperbolas. In each case we can choose new coordinates $(x, y, z) \in \mathbb{R}^3$ so that the slicing plane has equation $z = 1$. If we then homogenize as before, we end up with the original cone.

Already the $\mathbb{R}^3$ model has shown a remarkable power to unify. All ellipses, parabolas, and hyperbolas are simply different views of one and the same projective object. In any viewing screen, the conic section is an *affine curve*. The "global" view, the cone in $\mathbb{R}^3$ thought of as a curve in $\mathbb{P}^2(\mathbb{R})$, is a *projective curve*. ◊

Let's make some of these ideas official.

Definition B.7. In any real plane with coordinates (x, y), if the zero set of a polynomial $p(x, y)$ defines a curve C in $\mathbb{R}^2$, then C is called a *real affine plane curve* or an *affine curve* in $\mathbb{R}^2$. The homogenization $p(x, y, z)$ of $p(x, y)$ defines a homogeneous zero set in $\mathbb{R}^3$, and the 1-subspaces of

Further Examples

this set — points in $\mathbb{P}^2(\mathbb{R})$ — comprise a *real projective plane curve* or a *projective curve* in $\mathbb{P}^2(\mathbb{R})$. It is called the *projective completion* of C. ◊

Further Examples

In Figure B.5 the affine viewing screen is $z = 1$ and the points at infinity in it are the 1-subspaces of the plane $z = 0$. From the dimension-lowering observation made on p. 187, this plane, a 2-subspace of $\mathbb{R}^3$, corresponds to a projective line so all points at infinity form a line in $\mathbb{P}^2(\mathbb{R})$. The choice $z = 1$ was arbitrary. Any plane $ax + by + cz = 1$ can be regarded as a viewing plane, and its points at infinity are the lines through the origin in the 2-subspace $ax + by + cz = 0$. For every such choice of plane in $\mathbb{R}^3$, we determine a corresponding line at infinity. We can make qualitative observations about viewing planes more concrete by using equations.

Example B.8. For the cone in Figure B.5, we should be able to write, say, the equation of the hyperbola in which the plane $x = 1$ intersects the cone. Since we know the equation of the cone, this is easy: Substitute $x = 1$ into the cone's homogenized equation $(x - z)^2 + y^2 = 4z^2$ to get

$$(1 - z)^2 + y^2 = 4z^2.$$

For a more arbitrary plane parameterized by degree-one functions f, g, h as

$$\{x = f(u,v),\ y = g(u,v),\ z = h(u,v)\},$$

substitute $f(u,v)$, $g(u,v)$, $h(u,v)$ in for x, y, z in the cone's equation. ◊

Example B.9. Everything in the last few paragraphs directly generalizes to any algebraic curve C, giving us a mechanism for obtaining the equation of any algebraic curve in any viewing screen. Coupled with the formula for shrinking the plane to a disk, we can track views of C in the disk model of $\mathbb{P}^2(\mathbb{R})$ as the viewing plane changes. For example, the pictures in Figure B.6 depict rotated views of the 2×1 ellipse

$$\frac{x^2}{2^2} + \frac{y^2}{1^2} = 1.$$

Its homogenization $\frac{x^2}{2^2} + \frac{y^2}{1^2} = z^2$ defines an elliptical cone through the origin of $\mathbb{R}^3$, and the original ellipse sits in the plane $z = 1$. The fundamental

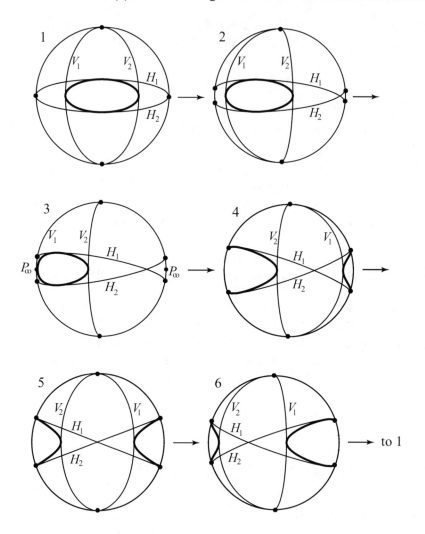

Figure B.6. View 1 shows a sphere with antipodal points identified, together with an ellipse and its horizontal and vertical tangents. The large bounding circle represents the line at infinity. In view 2, the sphere has rotated leftward but the line at infinity stays in place, so it has rotated relative to the ellipse. The rotation continues in the other views, and when view 6 goes to view 1, the amount of rotation is 180°, giving a horizontally flipped version of view 1. At 360°, double flipping gives exactly view 1.

Further Examples

rectangle surrounding the ellipse extends to the horizontal tangent lines H_1 and H_2 defined by $y = \pm 1$, and the vertical tangents V_1 and V_2 given by $x = \pm 2$. Our method allows us to follow the view of the ellipse and these four lines as the sphere rotates — or, equivalently, as the viewing plane, kept tangent to the unit sphere, slides along the sphere's equator. Figure B.6 shows six stages of this morphing. After the second time around, we arrive at the original view of the ellipse.

Since changing the viewing screen can take points at infinity and make them finite, we can get equations in any viewing screen, allowing us to analyze what happens around points that originally were at infinity. ◊

Appendix C

Some Code for This Book's Programs

This appendix contains suggested code for the programs in this book. Some are written in Maple, others in Mathematica, and importantly, some in the freeware GeoGebra which is easily available to all users.

Page 33: Creating animations using GeoGebra, Maple, or Mathematica.

GeoGebra:
In a browser, type GeoGebra and choose GeoGebra Classic. In the screen that appears, you'll see a blinking cursor in an input box. Enter into this box $x^2 - Bxy + y^2 - x - y = 0$. A slider for B appears, and by dragging the heavy dot, B varies and the conics simultaneously morph. GeoGebra is very intuitive, and by experimenting with what's in the box containing the slider, you can change slider limits, vary step size, ask the morphing to run continuously, and so on. You can also enter Discr= $B^2 - 4$ into a second input box GeoGebra creates after you filled in the first, and the discriminant's value is calculated on the fly.

GeoGebra can also easily handle several variables. For example, enter $Ax^2 + Bxy + Cy^2 = 1$ into the input box and GeoGebra creates an A-slider, a B-slider, and a C-slider. These sliders can be moved independently and GeoGebra will display the corresponding conic in real time.

Appendix C. Some Code for This Book's Programs

Recommendation: GeoGebra's online manual nicely explains the scope and capabilities of the software.

Maple code:
```
# This code animates the conic x^2+Bxy+y^2
# as B varies from -25 to 25.
plots[animate]( implicitplot, [x*x+B*x*y+y*y,x=-6..6,
  y=-6..6],B=-25..25);
```

Mathematica code:
```
(*This code animates the conic x^2+Bxy+y^2 *)
(*as B varies from -25 to 25. *)
  Animate[ContourPlot[x^2+B*x*y+y^2==1,{x,-6,6},
  {y,-6,6},PerformanceGoal->"Quality"],{B,-25,25}]
```

****

Page 33: Maple code finding triples (a, b, c) satisfying $p(a, b, c) = 0$.

For specificity, we take $p(a, b, c)$ below to be $-9a^2 + 7b^2 + ac = 41c^2$. However, the code works for the great majority of everyday cases.

```
#This code tests the triplets (a,b,c) in the box
#k determined by k and prints triples satisfying
#the equation.
k:=10;                              #Fix size of box.
for c from -k to k do
    for b from -k to k do
        for a from -k to k do
            if evalb(-9a^2+7b^2+ac=41c^2)=true
              then print (a,b,c);#Print good triples
            end if;
        od;
    od;
od;
```

In a small fraction of a second, the algorithm prints out the four points $(-1, 5, 2), (-1, -5, 2), (-2, 10, 4), (-2, -10, 4)$ on the curve

$$-9a^2 + 7b^2 + ac = 41c^2,$$

Appendix C. Some Code for This Book's Programs

with the last two points being multiples of the first two. Our program has found two rational points on the curve: $(x, y) = (\frac{-1}{2}, \pm\frac{5}{2})$. The discriminant of $-9x^2 + 7y^2 + x = 41$ is $-4 \cdot (-9) \cdot 7$ and is positive, so the curve is a hyperbola. Figure C.1 depicts the two rational points on the hyperbola.

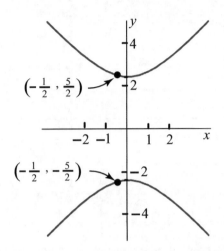

Figure C.1. The two rational points $(x, y) = (-\frac{1}{2}, \pm\frac{5}{2})$ are on the hyperbola $-9x^2 + 7y^2 + x = 41$. Choosing larger values for k increases the size of the search box and will yield other rational points lying outside the above plot-rectangle. In fact, in this case we can get not only rational points, but integer points. Examples are $(-12, \pm15)$ as well as $(22, \pm25)$.

Our search program can handle considerably more aggressive cases. Consider what happens when we just make up an equation with coefficients lying between ±30 like, say,

$$13a^2 + 25b^2 - 6ab + 17ac - 8bc = 11c^2$$

which can be translated to the curve

$$13x^2 + 25y^2 - 6xy + 17x - 8y = 11.$$

The discriminant of the curve's equation is negative, and there's an xy-term, so the curve is a tilted ellipse. Running the program with $k = 30$ and with $13a^2 + 25b^2 - 6ab + 17ac - 8bc = 11c^2$ in the "if" line quickly yields four integer triples (a, b, c) leading to four rational points on the ellipse. Two of them are $(-\frac{15}{11}, -\frac{7}{11})$ and $(-\frac{2}{19}, -\frac{11}{19})$. The other two are shown in Figure C.2.

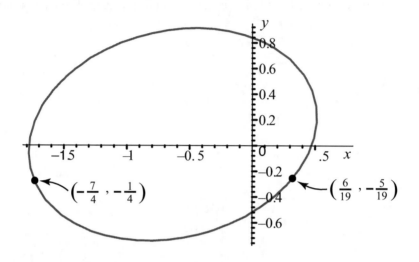

Figure C.2. Two rational points on a tilted ellipse whose equation has larger coefficients.

****

Page 46: Creating projective disk images using GeoGebra or Maple.

Parameterizing the parabola in $\mathbb{R}^2$ by

$$t \to (t, t^2 - c), \quad t \in \mathbb{R},$$

and dividing both x and $y - c$ by $\sqrt{x^2 + y^2 + 1}$ gives a parametrization of the disk image:

$$t \to \left(\frac{t}{\sqrt{t^2 + (t^2 - c)^2 + 1}}, \frac{t^2 - c}{\sqrt{t^2 + (t^2 - c)^2 + 1}} \right), \quad t \in \mathbb{R}.$$

Appendix C. Some Code for This Book's Programs

GeoGebra: Type the following into the GeoGebra input box:

(cos(t), sin(t))   **Creates the bounding circle**
T(x,y) = x/sqrt(x^2 + y^2 + 1)   **Defines the transformation for the disk image**
Curve(T(t, t^2), T(t^2, t), t, −10, 10)   **Plot the curve**

Note: Enter Curve(T(t, t^2 + b), T(t^2 + b, t), t, −10, 10) to create a slider. Vary it to see the effect.

Maple:

```
with(plots);
#p1 plot the bounding unit circle:
  p1:=plot([cos(t),sin(t),t=0..2*Pi],
    scaling=constrained, axes=none, color=black):
#p2 plot the x-axis:
  p2:=plot([t,0,t=-1..1],
    scaling=constrained, axes=none, color=black):
#p3 plot the y-axis:
  p3:=plot([0,t,t=-1..1],
    scaling=constrained, axes=none, color=black):
#p4 plot the image of the parabola:
  p4:=plot[(t/sqrt(t^2+t^4+1),t^2/sqrt(t^2+t^4+1),
    t=-10..10],scaling=constrained, axes=none,
      color=black):
#Plot the four components:
  display(p1,p2,p3,p4)
```

Notice the initial line with(plots); in the above Maple code. This activates Maple's plots library and lets the program use the display command to draw the four images p1,p2,p3,p4. Also, notice that each plot calculation ends with a colon instead of a semicolon to suppress filling up multiple screens of raw plotting data.

****

Page 77 : The connect-the-dots routine using GeoGebra Classic.

In GeoGebra Classic version 6, enter $y^2 = x^3 + bx + c$ into the input box and if GeoGebra Classic does not do it automatically, create sliders for b and c. To edit a slider, click on one of the endpoints. Click on the line icon and select two points on the curve. Use the Move Tool to move one or both of the points so you can see the third intersection point. Now locate the third intersection point: Enter `Intersect(eq1, f, p)` where `eq1` and `f` are the respective names GeoGebra Classic assigns to the implicit equation and the line found in the algebra pane. Edit the slider for p so that the range is `1..3` in increments of 1. Animate the slider for p, pausing when a point appears at the third intersection point. Note the name of the third intersection point in the algebra pane and type `Reflect(PointName,y=0)` in the input box. In the Settings menu you will find a grid of various colors. Right click on the new point and change its color so it becomes more easily recognizable. This point is the output of the connect-the-dots algorithm. You can now follow the connect-the-dots routine by noting the coordinates of the new dot in the algebra pane and moving one of the original dots to those coordinates. Move the p slider as needed to identify the third intersection point. Note that you cannot only vary the two points on the curve, but the curve itself can vary as the sliders for the coefficients b and c are moved. Although this routine does not provide exact values, it nicely illustrates the connect-the-dots method.

****

Page 77 : Philosophy for automating the connect-the-dots routine using Maple or Mathematica.

We outline here one approach to automating the connect-the-dots procedure using a symbolic manipulation package such as Maple or Mathematica.

To construct appropriate code, start with $y^2 = x^3 + ax + b$ as the curve's equation, with numeric values assigned a and b. Then, assuming we have a generator (x_1, y_1), assign the generator's coordinate values to x_1 and y_1. Here are the steps a computer could take so as to successively update points P_i in the connect-the-dots procedure.

Appendix C. Some Code for This Book's Programs

(a) The tangent line to the curve at $P_1 = (x_1, y_1)$ has slope
$$m = (3x_1^2 + a)/(2y_1).$$

(b) The tangent line's third intersection with the curve is
$$P_2 = P_1 + P_1 = (x_2, y_2) = \left(m^2 - 2x_1,\ -m(m^2 - 2x_1) - (y_1 - mx_1)\right).$$

(c) After initiating the process with the above tangent line, to get the next point $P_3 = P_2 + P_1 = (x_3, y_3)$ as the third intersection of the line L through (x_1, y_1) and the already-calculated point (x_2, y_2), observe that L's slope is now merely $(y_2 - y_1)/(x_2 - x_1)$. Obtain a formula for (x_3, y_3) that can be used to update input to successively get
$$P_4 = P_3 + P_1,\ P_5 = P_4 + P_1, \ldots.$$

Exercise C.1. Starting from $P_1 = (0, 4)$ on the curve $y^2 = x^3 - 4x + 16$, check that your code yields
$$P_4 = \left(\frac{-3{,}967}{15{,}376},\ \frac{7{,}864{,}639}{1{,}906{,}624}\right) \approx (-0.258, 4.125).$$

Bibliography

[Aitken] Wayne Aitken, *Math 372: Introduction to Number Theory — Supplement*, California State University at San Marcos, 2005.

[Apostol] Tom M. Apostol, *Introduction to analytic number theory*, Undergraduate Texts in Mathematics, Springer-Verlag, New York-Heidelberg, 1976. MR0434929

[Bhargava 1] Manjul Bhargava, *On the Conway-Schneeberger fifteen theorem*, Quadratic forms and their applications (Dublin, 1999), Contemp. Math., vol. 272, Amer. Math. Soc., Providence, RI, 2000, pp. 27–37, DOI 10.1090/conm/272/04395. MR1803359

[Bhargava 2] Manjul Bhargava, *Higher composition laws. I. A new view on Gauss composition, and quadratic generalizations*, Ann. of Math. (2) **159** (2004), no. 1, 217–250, DOI 10.4007/annals.2004.159.217. MR2051392

[Bhargava 3] Manjul Bhargava and Arul Shankar, *Binary quartic forms having bounded invariants, and the boundedness of the average rank of elliptic curves*, Ann. of Math. (2) **181** (2015), no. 1, 191–242, DOI 10.4007/annals.2015.181.1.3. MR3272925

[Bhargava 4] Manjul Bhargava and Arul Shankar, *Ternary cubic forms having bounded invariants, and the existence of a positive proportion of elliptic curves having rank 0*, Ann. of Math. (2) **181** (2015), no. 2, 587–621, DOI 10.4007/annals.2015.181.2.4. MR3275847

[Bhargava 5] Manjul Bhargava, *Most hyperelliptic curves over $\mathbb{Q}$ have no rational points*, http://arxiv.org/abs/1308.0395v1.

[Bhargava 6] Manjul Bhargava and Arul Shankar, *The average size of the 5-Selmer group of elliptic curves is 6, and the average rank is less than 1*, http://arxiv.org/abs/1312.7859v1.

[BSZ] Manjul Bhargava, Christopher Skinner, and Wei Zhang, *A majority of elliptic curves over $\mathbb{Q}$ satisfy the Birch and Swinnerton-Dyer conjecture*, http://arxiv.org/abs/1407.1826v2.

[BHKSSW] Jennifer S. Balakrishnan, Wei Ho, Nathan Kaplan, Simon Spicer, William Stein, and James Weigandt, *Databases of elliptic curves ordered by height and distributions of Selmer groups and ranks*, LMS J. Comput. Math. **19** (2016), no. suppl. A, 351–370, DOI 10.1112/S1461157016000152. MR3540965

[Cremona] J. E. Cremona and D. Rusin, *Efficient solution of rational conics*, Math. Comp. **72** (2003), no. 243, 1417–1441, DOI 10.1090/S0025-5718-02-01480-1. MR1972744

Bibliography

[Fischer] Gerd Fischer, *Plane algebraic curves*, translated from the 1994 German original by Leslie Kay, Student Mathematical Library, vol. 15, American Mathematical Society, Providence, RI, 2001. MR1836037

[Fuji-Oike] Kazuyuki Fuji and Hiroshi Oike, *An algebraic proof of the associative law of elliptic curves*, Advances in Pure Mathematics, Volume 7, pp. 649–659, 2017.

[Griffiths] Phillip A. Griffiths, *Introduction to algebraic curves*, translated from the Chinese by Kuniko Weltin, Translations of Mathematical Monographs, vol. 76, American Mathematical Society, Providence, RI, 1989. MR1013999

[Hindry] Marc Hindry and Joseph H. Silverman, *Diophantine geometry: An introduction*, Graduate Texts in Mathematics, vol. 201, Springer-Verlag, New York, 2000. MR1745599

[Hungerford] Thomas W. Hungerford, *Algebra*, reprint of the 1974 original, Graduate Texts in Mathematics, vol. 73, Springer-Verlag, New York-Berlin, 1980. MR600654

[Husemöller] Dale Husemoller, *Elliptic curves*, with an appendix by Ruth Lawrence, Graduate Texts in Mathematics, vol. 111, Springer-Verlag, New York, 1987. MR868861

[Ireland] Kenneth Ireland and Michael Rosen, *A classical introduction to modern number theory*, 2nd ed., Graduate Texts in Mathematics, vol. 84, Springer-Verlag, New York, 1990. MR1070716

[Kendig 1] Keith Kendig, *Conics*, with 1 CD-ROM (Windows NT/2000, XP, Macintosh), The Dolciani Mathematical Expositions, vol. 29, Mathematical Association of America, Washington, DC, 2005. MR2146621

[Kendig 2] Keith Kendig, *A guide to plane algebraic curves*, The Dolciani Mathematical Expositions, vol. 46, MAA Guides, 7, Mathematical Association of America, Washington, DC, 2011. MR2815937

[Kendig 3] Keith Kendig, *Elementary algebraic geometry*, 2nd ed., revision of Springer-Verlag, Graduate Texts in Mathematics, Vol. 44, Dover Publications, Inc., 2015.

[LMFDB] The LMFDB Collaboration, *The L-functions and modular forms database*, http://www.lmfdb.org, 2013.

[Massey] William S. Massey, *Algebraic topology: An introduction*, reprint of the 1967 edition, Graduate Texts in Mathematics, Vol. 56, Springer-Verlag, New York-Heidelberg, 1977. MR0448331

[Mazur] B. Mazur, *Rational isogenies of prime degree (with an appendix by D. Goldfeld)*, Invent. Math. **44** (1978), no. 2, 129–162, DOI 10.1007/BF01390348. MR482230

[PPVW] Jennifer Park, Bjorn Poonen, John Voight, and Melanie Matchett Wood, *A heuristic for boundedness of ranks of elliptic curves*, J. Eur. Math. Soc. (JEMS) **21** (2019), no. 9, 2859–2903, DOI 10.4171/JEMS/893. MR3985613

[Serre] J.-P. Serre, *A course in arithmetic*, translated from the French, Graduate Texts in Mathematics, No. 7, Springer-Verlag, New York-Heidelberg, 1973. MR0344216

[Silverman] Joseph H. Silverman and John Tate, *Rational points on elliptic curves*, Undergraduate Texts in Mathematics, Springer-Verlag, New York, 1992. MR1171452

Bibliography

[Stoll] Douglas A. Stoll and Patrick Demichel, *The impact of $\zeta(s)$ complex zeros on $\pi(x)$ for $x < 10^{10^{13}}$*, Math. Comp. **80** (2011), no. 276, 2381–2394, DOI 10.1090/S0025-5718-2011-02477-4. MR2813366

[Tanaka] Minoru Tanaka, *A numerical investigation on cumulative sum of the Liouville function*, Tokyo J. Math. **3** (1980), no. 1, 187–189, DOI 10.3836/tjm/1270216093. MR584557

[Whitney] Hassler Whitney, *Complex analytic varieties*, Addison-Wesley Publishing Co., Reading, Mass.-London-Don Mills, Ont., 1972. MR0387634

Index

15 Theorem, 134, 135
290 Theorem, 134

affine, 53, 55
affine curve, 189
affine curve in $\mathbb{R}^2$, 188
Aitken, Wayne, 34, 101, 167, 201
Albers, Donald, xiv
alpha curve, 62, 63, 183
Apostol, Tom, 120, 201
Arithmetica, 134

Balakrishnan, Jennifer, 201
Bhargava, Manjul, 127, 128, 130–135, 160, 201
Bhargava, Mira, 130
Birch–Swinnerton-Dyer Conjecture, 93, 102, 105, 106, 111, 113, 125, 128, 129, 134
Brumer, Armand, 125
BSD, 127, 129

circle, 3, 4, 7–9, 11–13, 21, 31–33, 41, 53, 138–142, 144, 148, 155
Clay Institute, 93, 116, 125
complex, 138–143, 145, 146, 148, 150, 151, 157–159
conductor, 123
conic, 16, 21, 22, 30–33, 60, 73, 161
connect-the-dots, 77, 108, 118, 145, 146
Conway, John H., 135
coset, 147–150
Cremona, John, 33, 34, 101, 167, 201
crosspoint, 71

cubic, 50, 55–58, 63, 64, 69, 74, 77, 78, 86, 146, 158
cusp, 63, 69, 71, 119
cyclic, 84, 85, 88, 89, 91, 92, 96, 100, 148

dehomogenization, 52, 62
Demichel, Patrick, 124, 203
Diophantine equation, xiii, 16
Diophantus, 132, 134
discriminant, 30–32, 68, 69, 71, 92, 119, 195
disk model, 42, 53, 74, 85, 140
disk model of $\mathbb{P}^2(\mathbb{R})$, 183

Elkies, Noam, 94
ellipse, 17, 18, 24, 25, 30–32, 38, 59, 60, 196
elliptic curve, 55, 57–60, 63–67, 69–72, 75, 77–79, 81, 82, 84, 86–88, 90, 92–97, 100, 101, 105, 107, 110, 111, 116–122, 127–129, 134, 137, 139, 142, 143, 145, 148, 149, 152, 158, 161, 169, 174
Euclid, 6, 8
expected rank, 119, 120
expected value, 111

Faltings's Theorem, 159, 161
Faltings, Gerd, 121, 161
Fermat's Last Theorem, 3, 56, 87, 170
Fermat, Pierre, 3, 4, 56, 89, 132, 134
Fields Medal, 134, 161
finitely generated abelian group, 92
Fischer, Gerd, 202
Four Squares Theorem, 134

Fuji, Kazuyuki, 202

Gauss, Karl, 132–134
genus, 137, 138, 156–161
GeoGebra, xi, 33, 46, 193, 197, 198
Granville, Andrew, 135
GRH, 127
Griffiths, Phillip, 202
group, 83–92, 94–100, 145–150

Hadamard, Jacques, 161
Hanke, Jonathan, 134
Heath-Brown, Roger, 125
height, 120, 121, 123, 127
hemisphere model of $\mathbb{P}^2(\mathbb{R})$, 185
Hindry, Marc, 202
Ho, Wei, 201
homogeneous, xiii, 16, 38, 52, 72, 77, 102, 137, 187, 188
homogenization, 49, 52, 60, 187
homogenization of a set, 187
Hungerford, Thomas, 92, 202
Husemöller, D., 80, 202
hyperbola, 22, 23, 38, 39, 45, 55, 140–142, 195

identity component, 66
Implicit Function Theorem, 172
inflection point, 48, 84, 87
Ireland, Kenneth, 11, 111, 202
irreducible, 58–60, 158

Kaplan, Nathan, 201
Katz, Nicholas, 124
Kendig, Keith, 30, 158, 164, 202
Kennedy, Stephen, xiv

Lagrange, Joseph-Louis, 134
lattice, 119, 120, 127, 129, 148
Legendre's Criterion, 11, 33, 34
Legendre, Adrien-Marie, 10, 33, 34
LMFDB, 90, 93, 95, 97, 100, 102, 103, 170, 202
long form, 65, 89, 90

Maple, xi, 33, 46, 108, 193, 194, 196, 197
Massey, W. S., 156, 202
Mathematica, xi, 46, 193, 194

Mazur's Torsion Theorem, 95, 96, 161
Mazur, Barry, 94–96, 161, 202
Mordell's Theorem, 80, 94
Mordell, Louis, 80, 94, 161

natural numbers, 29
nondegenerate, 16, 20–22, 57, 58, 72, 137, 160, 161
nonsingular, 159
nonsingular curve, 61, 174

Oike, Hiroshi, 202
order, 85, 92, 96, 100, 102, 103, 148, 177
Ore, Oystein, 94
oval, 51, 60, 66, 71, 146–149

parabola, 26–28, 30, 31, 38, 39, 43–45, 55, 60, 142, 161, 196
Park, Jennifer, 202
plane algebraic curve, 2
Plimpton 322, 5
Plimpton, George, 5
point at infinity, 43, 48, 51, 76, 81, 84, 86, 89, 90, 97, 146
Pollatsek, Harriet, xiv
Polya's Conjecture, 124
Poonen, Bjorn, 202
prime, 21, 91, 119, 124, 132
Prime Number Theorem, 119
primitive solution, 14, 15, 19, 20, 23, 24, 26, 29, 35
projective curve in $\mathbb{P}^2(\mathbb{R})$, 189
projective plane, 42, 44, 53, 143
Pythagorean Theorem, 4, 5, 132
Pythagorean triple, 6

quadratic, 8, 12, 15, 17, 22, 30, 55, 72, 73, 133, 134

Ramanujan, 134
rank, 93–95, 97–101, 105, 106, 116–120, 122, 123, 125, 127–129
rational circle, 7–9, 12, 21
rational conic, 72
rational curve, 2, 16, 55, 56, 78
rational ellipse, 25
rational hyperbola, 22
rational plane, 3, 7

Index

rational point, 3, 8–11, 16–18, 22, 23, 33–35, 55–57, 59, 60, 67, 73, 74, 76, 78–88, 90, 94, 95, 102, 105, 116, 120, 129, 159–161, 196
rational quadratic, 22
real affine plane curve, 188
real projective curve, 189
reflect, 72–75, 79, 80, 84–88, 97, 143, 144, 148, 149, 152
right triangle, 14, 132
Rosen, Michael, 202
Rosentrater, Ray, xiv
Rubik's Cube, 134
Rubik's minicube, 133
Rubinstein, Michael, 124
Rusin, D., 34, 201

Sarnak, Peter, 124, 135
Serre, Jean-Pierre, 34, 101, 202
Shankar, Arul, 201
short form, 65, 67, 70, 119
Silverman, Joseph, 64, 84, 202
singularity, 63, 70
Skinner, Christopher, 128, 201
smooth, 58, 61, 62, 64, 68, 71, 72, 74, 159
Spicer, Simon, 201
Stein, William, 201
Stoll, Douglas, 124, 203
structure of finitely generated abelian groups, 92
subgroup, 147, 148, 150

Tanaka, Minoru, 124, 203
tangent, 45, 73, 74, 76, 84–86, 88, 97, 146, 199
Tate, John, 202
topological, 44, 53, 58, 148, 150, 158
topology, 2, 53, 58
topology of $\mathbb{P}^2(\mathbb{R})$, 186
torsion, 94–96, 99–101, 103, 120, 161
transversal, 48

vector space model of $\mathbb{P}^2(\mathbb{R})$, 186
Voight, John, 202

Weierstrass form, 64, 65
Weierstrass long form, 89, 168
Weierstrass short form, 66, 70, 73, 89, 119, 168, 169
Weierstrass, Karl, 65–67, 73, 78, 86, 89, 90, 94, 100, 119, 169
Weigandt, James, 201
Weil, André, 161
Whitney, Hassler, 203
Wiles, Andrew, 3
Wood, Malanie, 202

Young, Matthew, 125

Zhang, Wei, 128, 201